Student Success Organizer

PRECALCULUS FUNCTIONS AND GRAPHS: A GRAPHING APPROACH

THIRD EDITION

and

PRECALCULUS WITH LIMITS: A GRAPHING APPROACH

THIRD EDITION

Larson/Hostetler/Edwards

Emily J. Keaton

HOUGHTON MIFFLIN COMPANY **Boston** **New York**

Editor-in-Chief: Jack Shira
Managing Editor: Cathy Cantin
Development Manager: Maureen Ross
Associate Editor: Laura Wheel
Assistant Editor: Rosalind Horn
Supervising Editor: Karen Carter
Project Editor: Patty Bergin
Editorial Assistant: Lindsey Gulden
Production Technology Supervisor: Gary Crespo
Marketing Manager: Michael Busnach
Marketing Assistant: Nicole Mollica
Senior Manufacturing Coordinator: Sally Culler

Printed in the United States of America

ISBN 0-618-09845-3

123456789-VGI-05 04 03 02 01

CONTENTS

TO THE STUDENT

The Student Success Organizer for Precalculus Functions & Graphs: A Graphing Approach, Third Edition, and for Precalculus with Limits: A Graphing Approach, Third Edition, is a supplement to these texts by Ron Larson, Robert P. Hostetler and Bruce H. Edwards. All references to chapters, sections, theorems, definitions and examples apply to these texts.

This notebook organizer will help you develop a section-by-section summary of the key concepts in your textbook to better study and prepare for exams. It is a set of templates designed to help you take notes, review section highlights, draw graphs, and work on examples. Special boxes help you review vocabulary, track section objectives, and record homework assignments.

How to Use the Student Success Organizer

Each section of the Student Success Organizer corresponds to a text section, allowing you to study the sections assigned by your instructor. The Student Success Organizer was designed to be flexible, so you can use the material in the way that best fits your style of studying. Some suggestions for using the Student Success Organizer in class preparation, note taking, studying, and reviewing for exams are given below.

Using the Student Success Organizer to prepare for class:
- In the text, read the section you will cover in class.
- Work through each part of the Student Success Organizer for that section, filling in blanks and working each example.
- Make a note of questions you have about a topic so you can ask about it in class or during your instructor's office hours.

Using the Student Success Organizer during class:
- Bring the sections you will cover that day to class with you.
- As the instructor covers each part of the material, take notes directly on the Student Success Organizer about vocabulary, concepts, and examples. Use this to supplement the notes you take in your regular notebook.
- When the instructor finishes covering a topic, check your notes to make sure all your questions about that topic are answered. Ask questions about any concept you do not understand before the instructor moves on to the next topic.
- Write down the assigned homework at the end of each section of the Student Success Organizer.

Using the Student Success Organizer to review and prepare for exams:
- After each class, review your notes in your notebook and the Student Success Organizer and fill in any gaps. If you still have questions, make a note of them to ask your instructor in class or during office hours. These gaps can also help you identify areas for further review.
- Read your notes in your notebook and in the Student Success Organizer before exams to make sure you understand key vocabulary, concepts, and formulas.
- If you need further practice working on a topic, check the Homework Assignment for that section to see which exercises you have already finished. This will allow you to work through review exercises for that section which you have not yet tried.

STUDY STRATEGIES

Your success in mathematics depends on your active participation both in class and outside of class. Because the material you learn each day builds on the material you learned previously, it is important that you keep up with your course work every day and develop a clear plan of study. This set of guidelines highlights key study strategies to help you learn how to study mathematics.

Preparing for Class

The syllabus your instructor provides is an invaluable resource that outlines the major topics to be covered in the course. Use it to help you prepare. As a general rule, you should set aside two to four hours of study time for each hour spent in class. Being prepared is the first step toward success. Before class:

- Review your notes from the previous class.
- Read the portion of the text that will be covered in class.
- Use the objectives listed at the beginning of each section to keep you focused on the main ideas of the section.
- Pay special attention to the definitions, rules and concepts highlighted in boxes. Also, be sure you understand the meanings of mathematical symbols and terms written in boldface type. Keep a vocabulary journal for easy reference.
- Read through the solved examples. Use the side comments given in the solution steps to help you in the solution process. Also, read the *Study Tips* given in the margins.
- Make notes of anything you do not understand as you read through the text. If you still do not understand after your instructor covers the topic in question, ask questions before your instructor moves on to a new topic.

Keeping Up

Another important step toward success in mathematics involves your ability to keep up with the work. It is very easy to fall behind, especially if you miss a class. To keep up with the course work, be sure to:

- Attend every class. Bring your text, a notebook, a pen or pencil, and a calculator (scientific or graphing). If you miss a class, get the notes from a classmate as soon as possible and review them carefully.
- Participate in class. As mentioned above, if there is a topic you do not understand, ask about it before the instructor moves on to a new topic.
- Take notes in class. After class, read through your notes and add explanations so that your notes make sense to *you*. Fill in any gaps and note any questions you might have.
- Reread the portion of your text that was covered in class. This time, work each example *before* reading through the solution.
- Do your homework as soon as possible, while concepts are still fresh in our mind. Allow at least two hours outside of each class for homework so you do not fall behind. Learning mathematics is a step-by-step process, and you must understand each topic in order to learn the next one.
- Use your notes from class, the text discussion, the examples, and the *Study Tips* as you do your homework. Many exercises are keyed to specific examples in the text for easy reference.

- When you are working problems for homework assignments, show every step in your solution. Then, if you make an error, it will be easier to find where the error occurred.

Getting Extra Help

It can be very frustrating when you do not understand concepts and are unable to complete homework assignments. However, there are many resources available to help you with your studies.

- Your instructor may have office hours. If you are feeling overwhelmed and need help, make an appointment to discuss your difficulties with your instructor.
- Find a study partner or a study group. Sometimes it helps to work through problems with another person.
- Arrange to get regular assistance from a tutor. Many colleges have a math resource center available on campus, as well.
- Consult one of the many ancillaries available with this text: the Student Solutions Guide, tutorial software, videotapes, and additional study resources available at our website at *www.hmco.com*.

Preparing for an Exam

The last step toward success in mathematics lies in how you prepare for and complete exams. If you have followed the suggestions given above, then you are almost ready for exams. Do not assume that you can cram for the exam the night before—this seldom works. As a final preparation for the exam,

- Read the *Chapter Summary*, keyed to each section, and review the concepts and terms.
- Work through the *Chapter Review Exercises* if you need extra practice on material from a particular section. You can practice for an exam by first trying to work through the exercises with your book and notebook closed.
- Take the *Chapter Test* as if you were in class. You should set aside at least one hour per test. Check your answers against the answers given in the back of the book. Work these problems a few days before the exam and review any areas of weakness.
- Review your notes and the portion of the text that will be covered on the exam.
- Start studying for your exam well in advance (at least a week). The first day or two, only study about two hours. Gradually increase the study time each day. Be completely prepared for the exam two days in advance. Spend the final day just building confidence so you can be relaxed during the exam.
- When you study for an exam, first look at all definitions, properties, and formulas until you know them. Then work as many exercises as you can, especially any kind of exercise that has given you trouble in the past.
- Avoid studying up until the last minute. This will only make you anxious. Allow yourself plenty of time to get to the testing location.
- Once the exam begins, read through the directions and the entire exam before beginning. Work the problems that you know how to do first to avoid spending too much time of the exam on any one problem. Time management is extremely important when taking an exam.
- If you finish early, use the remaining exam time to go over your work.
- When you get an exam back, review it carefully and go over your errors. Rework the problems you answered incorrectly. Discovering the mistakes you made will help you improve your test-taking ability.

Chapter P Prerequisites

Course Number

Instructor

Date

Section P.1 Graphical Representation of Data

Objective: In this lesson you learned how to plot points in the coordinate plane and use the Distance and Midpoint Formulas.

Important Vocabulary Define each term or concept.

Rectangular coordinate system

Cartesian plane

I. The Cartesian Plane (Pages 2–3)

What you should learn
How to plot points in the Cartesian plane

An ordered pair is . . .

On the Cartesian plane, the horizontal real number line is usually called the , and the vertical real number line is usually called the . The origin is the of these two axes, and the two axes divide the plane into four parts called

On the Cartesian plane shown below, label the x-axis, the y-axis, the origin, Quadrant I, Quadrant II, Quadrant III, and Quadrant IV.

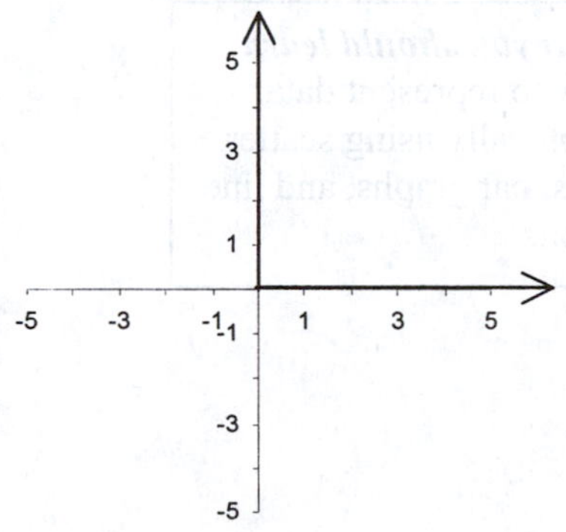

Larson/Hostetler/Edwards *Precalculus Functions and Graphs: A Graphing Approach, Third Edition*
Larson/Hostetler/Edwards *Precalculus with Limits: A Graphing Approach, Third Edition*
Student Success Organizer

Example 1: Explain how to plot the ordered pair (3, – 2), and then plot it on the Cartesian plane provided.

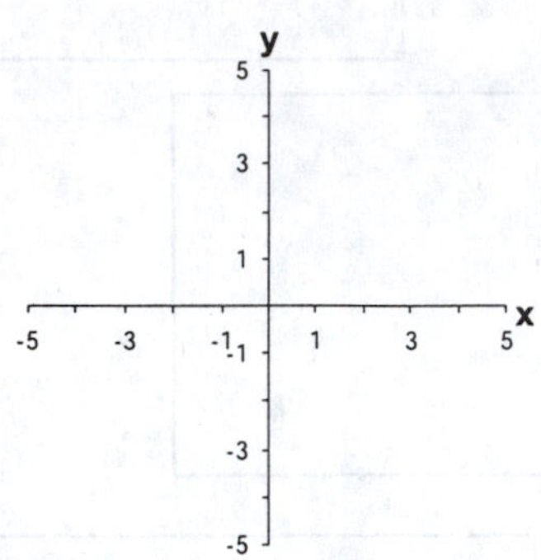

To shift a figure plotted in the rectangular coordinate system by *a* units to the left and *b* units upward, . . .

If (x, y) is an original point on a graph, is a reflection of this original point in the *y*-axis. If (x, y) is an original point on a graph, is a reflection of the original point in the *x*-axis. If (x, y) is an original point, is a reflection of the original point through the origin.

II. Representing Data Graphically (Pages 4–5)

To sketch a scatter plot of paired data given in a table, . . .

> ***What you should learn***
> How to represent data graphically using scatter plots, bar graphs, and line graphs

To create a bar graph of paired data given in a table, . . .

To create a line graph of paired data given in a table, . . .

III. The Distance Formula (Pages 5–6)

> ***What you should learn***
> How to use the Distance Formula to find the distance between two points

The **Distance Formula** states that . . .

Example 2: Explain how to use the Distance Formula to find the distance between the points (4, 2) and (5, – 1). Then find the distance and round to the nearest hundredth.

Example 3: Explain how to use a graphical solution to find the distance between the points (4, 2) and (5, – 1).

IV. The Midpoint Formula (Page 7)

> ***What you should learn***
> How to use the Midpoint Formula to find the midpoint of a line segment

The **midpoint** of a line segment is the point that subdivides the segment into two portions of length.

The **Midpoint Formula** gives the midpoint of the segment joining the points (x_1, y_1) and (x_2, y_2) as . . .

Example 4: Explain how to find the midpoint of the line segment with endpoints at (– 8, 2) and (6, – 10). Then find the coordinates of the midpoint.

What you should learn
How to find the equation of a circle

V. The Equation of a Circle (Page 8)

A **circle of radius *r*** in the plane consists of . . .

The **standard form of the equation of a circle** with center (h, k) and radius r is ______________________.

The standard form of the equation of a circle with radius r and its center at the origin is ______________________.

Example 5: For the equation $(x+2)^2 + (y-1)^2 = 4$, find the center and radius of the circle and then sketch the graph of the equation.

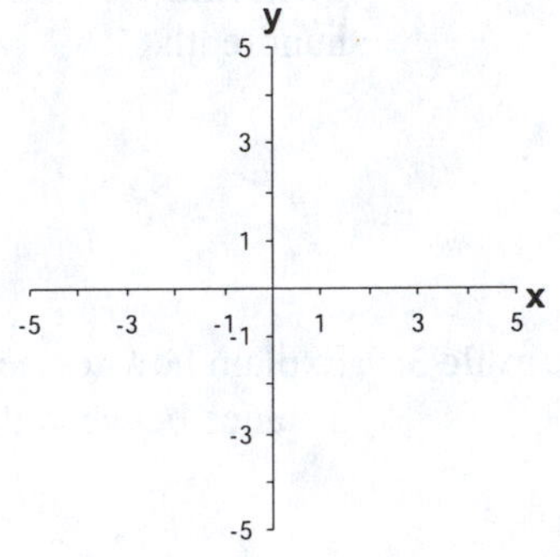

Additional notes

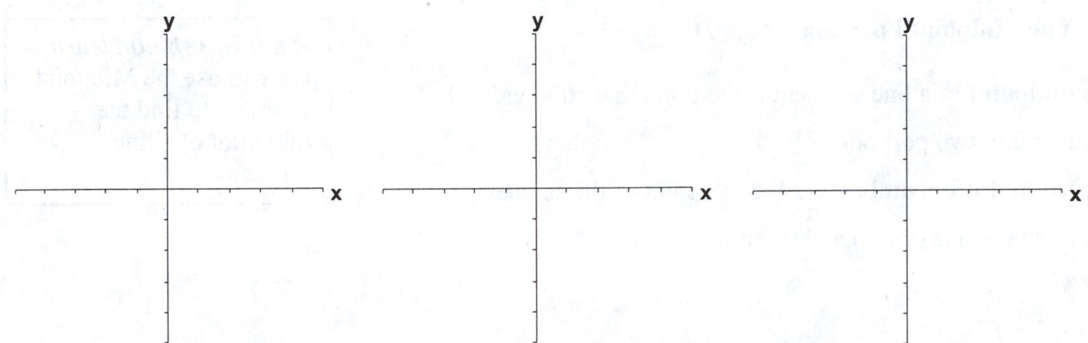

Homework Assignment

Page(s)

Exercises

Course Number
Instructor
Date

Section P.2 Graphs of Equations

Objective: In this lesson you learned how to sketch graphs of equations by point plotting or using a graphing utility.

Important Vocabulary	Define each term or concept.
Solution point	
Graph of an equation	
Intercepts	

I. The Graph of an Equation (Pages 14–15)

What you should learn How to sketch graphs of equations by point plotting

To sketch the graph of an equation by point plotting, . . .

Example 1: Complete the table for the equation $y = 3 - 0.5x$. Then use point plotting to sketch the graph of the equation.

x	-4	-2	0	2	4
y					

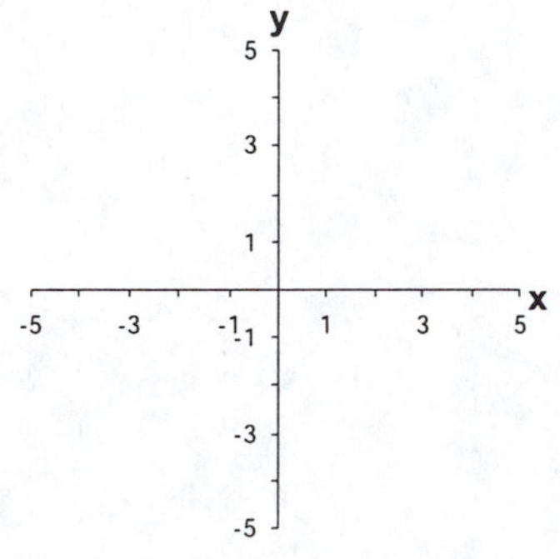

II. Using a Graphing Utility (Pages 16–19)

What you should learn
How to sketch graphs of equations using a graphing utility

A disadvantage of the point-plotting method is . . .

To graph an equation involving x and y on a graphing utility, . . .

Example 2: Use a graphing utility to graph the equation $12x^2 + 4y = 5$ in a standard viewing window.

A square setting is . . .

A square setting is useful when using a graphing utility to graph . . .

Example 3: Describe how to use a graphing utility to graph $3x^2 + 3y^2 = 75$. Then graph the equation in a square viewing window.

List and describe three common approaches to solving a problem.

1)

2)

3)

III. Applications of Graphs of Equations (Pages 19–20)

What you should learn
How to use graphs of equations in real-life problems

Describe a real-life situation in which a graphical solution approach would be helpful.

Example 4: Suppose a toy company estimates that its top-selling toy sells 240 units per minute, on average, nationally during the holiday shopping season, or according to the equation $S = 240m$, where S is the number of units sold and m is the number of minutes. Explain how a graphing utility could be used to find how long it takes during the holiday shopping season to sell 82,800 units.

Additional notes

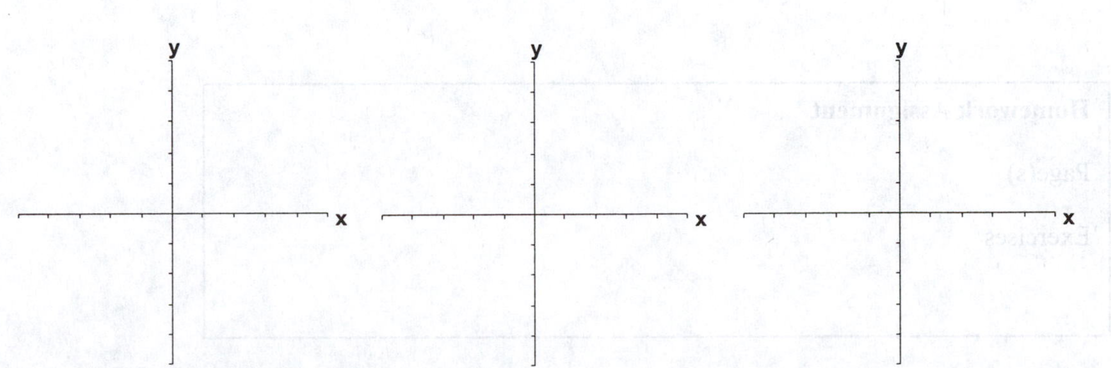

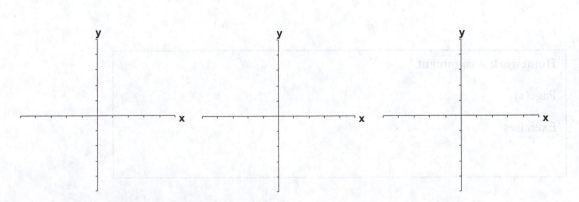

Additional notes

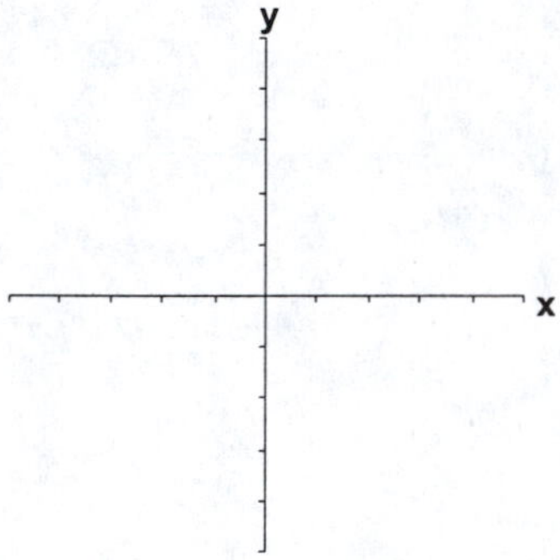

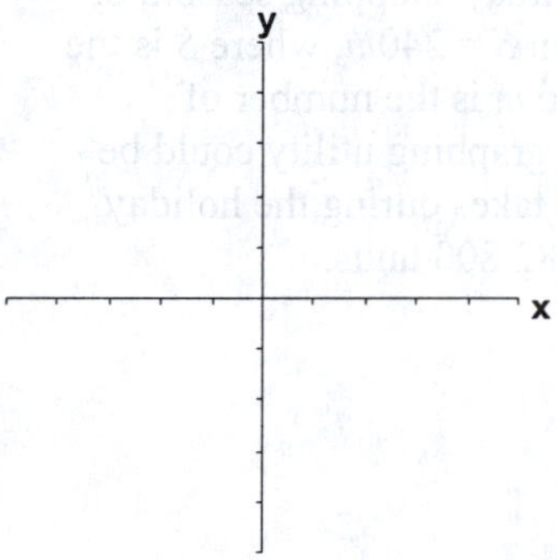

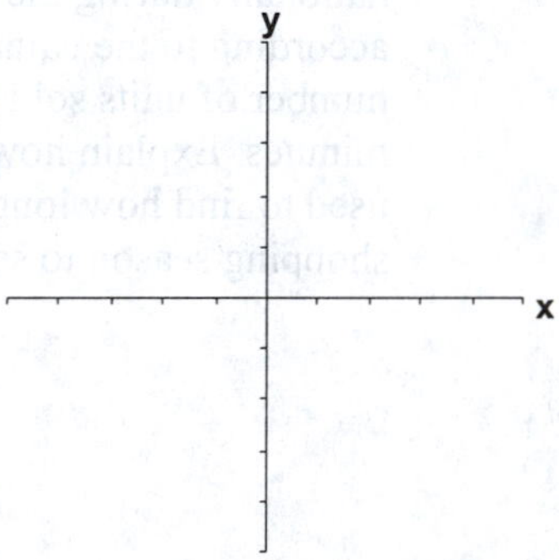

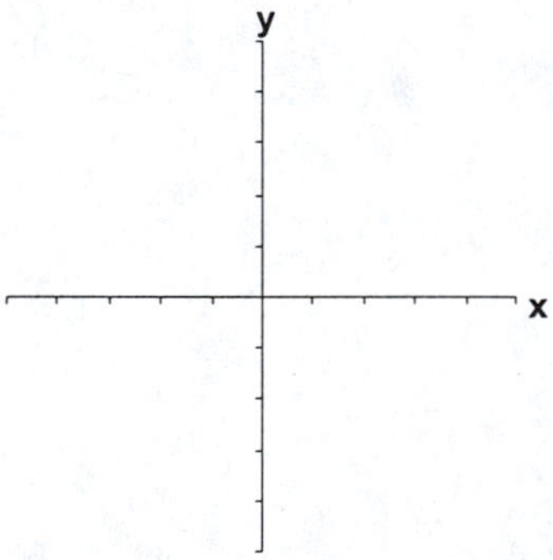

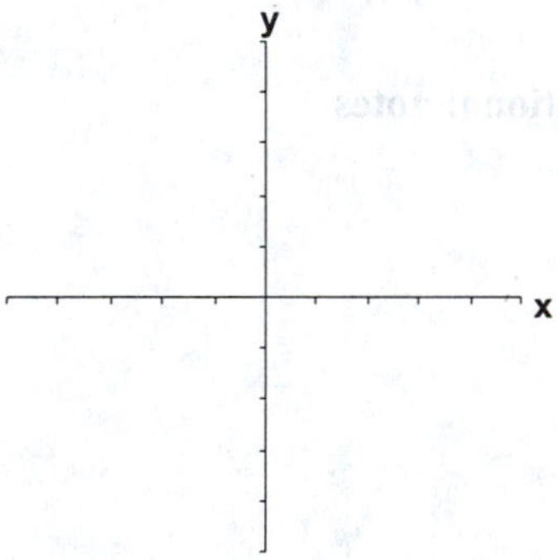

y x y x y x

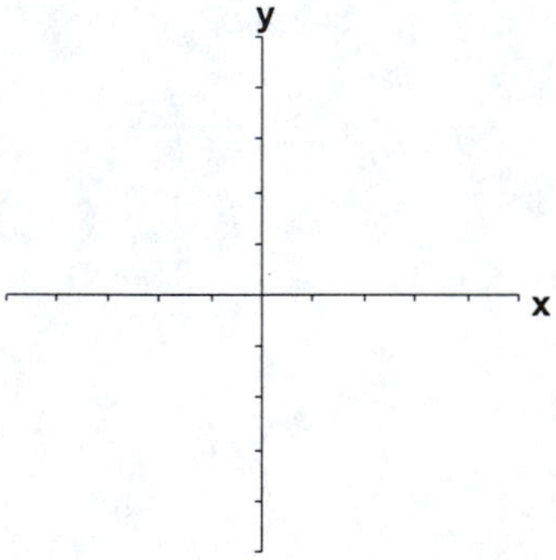

y x y x y x

Homework Assignment

Page(s)

Exercises

Course Number

Instructor

Date

Section P.3 Lines in the Plane

Objective: In this lesson you learned how to find and use the slope of a line to write and graph linear equations.

Important Vocabulary Define each term or concept.

Slope

Parallel

Perpendicular

I. The Slope of a Line (Pages 25–26)

What you should learn
How to find the slopes of lines

The formula for the **slope** of a line passing through the points (x_1, y_1) and (x_2, y_2) is $m =$

To find the slope of the line through the points $(-2, 5)$ and $(4, -3)$, . . .

A line whose slope is positive from left to right.

A line whose slope is negative from left to right.

A line with zero slope is

A line with undefined slope is

II. The Point-Slope Form of the Equation of a Line (Pages 27–28)

What you should learn
How to write linear equations given points on lines and their slopes

The **point-slope form** of the equation of a line is

This form of equation is best used to find the equation of a line when . . .

The **two-point form** of the equation of a line is

The two-point form of equation is best used to find the equation of a line when . . .

Example 1: Find an equation of the line having slope – 2 that passes through the point (1, 5).

The approximation method used to estimate a point between two given points is called . The approximation method used to estimate a point lying outside the given points is called

III. Sketching Graphs of Lines (Pages 29–30)

The **slope-intercept form** of the equation of a line is , where *m* is the and the *y*-intercept is (,).

What you should learn
How to use slope-intercept forms of linear equations to sketch graphs of lines

Example 2: Determine the slope and *y*-intercept of the linear equation $2x - y = 4$.

The equation of a **horizontal line** is . The slope of a horizontal line is . The *y*-coordinate of every point on the graph of a horizontal line is .

The equation of a **vertical line** is . The slope of a vertical line is . The *x*-coordinate of every point on the graph of a vertical line is .

The **general form** of the equation of a line is

.

Every line has an equation that can be written in

.

When a graphing utility is used to sketch a straight line, the graph of the line may not visually appear to have the slope indicated by its equation because . . .

In general, two graphs of the same equation can appear to be quite different depending on . . .

Example 3: Use a graphing utility to graph the linear equation $2x - y = 4$ using (a) a standard viewing window, and (b) a square window.

IV. Parallel and Perpendicular Lines (Pages 31–32)

What you should learn
How to use slope to identify parallel and perpendicular lines

Two lines are ________________ if they do not intersect.

Two lines are________________ if they intersect at right angles.

The relationship between the slopes of two lines that are parallel is . . .

The relationship between the slopes of two lines that are perpendicular is . . .

A line that is parallel to a line whose slope is 2 has slope ______.

A line that is perpendicular to a line whose slope is 2 has slope

_______.

Example 4: Use a graphing utility to graph the perpendicular lines $y = 2x - 3$ and $y = -0.5x + 5$ using (a) a standard viewing window, and (b) a square window.

Additional notes

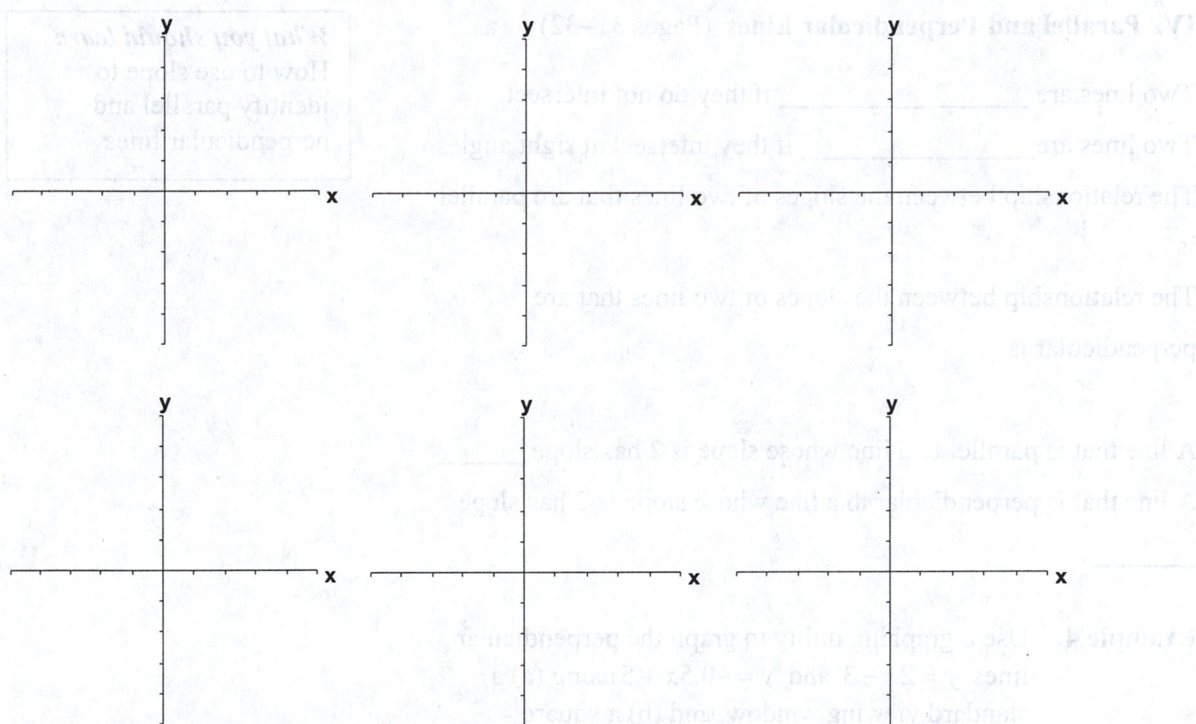

Homework Assignment

Page(s)

Exercises

Course Number

Instructor

Date

Section P.4 Solving Equations Algebraically and Graphically

Objective: In this lesson you learned how to solve linear equations, quadratic equations, polynomial equations, equations involving radicals, equations involving fractions, and equations involving absolute values.

Important Vocabulary	Define each term or concept.
Equation	
Extraneous	
***x*-intercept**	
***y*-intercept**	
Point of intersection	

I. Equations and Solutions of Equations (Pages 38–39)

What you should learn
How to solve linear equations

To solve an equation in x means to . . .

The values of x for which the equation is true are called its

_____ _____.

An identity equation is . . .

A conditional equation is . . .

A **linear equation in one variable x** is an equation that can be written in the standard form ____ ____ , where a and b are real numbers with $a \neq$ ___ ___.

Example 1: Solve $5(x + 3) = 35$.

To solve an equation involving fractional expressions, . . .

Larson/Hostetler/Edwards *Precalculus Functions and Graphs: A Graphing Approach, Third Edition*
Larson/Hostetler/Edwards *Precalculus with Limits: A Graphing Approach, Third Edition*
Student Success Organizer

When is it possible to introduce an extraneous solution?

Example 2: Solve: (a) $\frac{5x}{7}=\frac{9}{14}$ (b) $\frac{1}{x+1}+\frac{5x}{x^2-1}=\frac{4}{x-1}$

II. Intercepts and Solutions (Pages 39–41)

> ***What you should learn***
> How to find x- and y-intercepts of graphs of equations

To find the x-intercepts of the graph of an equation, . . .

To find the y-intercepts of the graph of an equation, . . .

Example 3: For the equation $3x-4y=12$, find:
(a) the x-intercept(s), and (b) the y-intercept(s).

III. Finding Solutions Graphically (Pages 41–42)

> ***What you should learn***
> How to find solutions of equations graphically

To use a graphing utility to graphically approximate the solutions of an equation , . . .

Example 4: Use a graphing utility to approximate the solutions of $3x^2-14x=-8$.

IV. Points of Intersection of Two Graphs (Pages 43–44)

> ***What you should learn***
> How to find the points of intersection of two graphs

To find the points of intersection of the graphs of two equations algebraically, . . .

To find the points of intersection of the graphs of two equations with a graphing utility, . . .

Example 5: Use (a) an algebraic approach and (b) a graphical approach to finding the points of intersection of the graphs of $y = 2x^2 - 5x + 6$ and $x - y = -6$.

V. Solving Polynomial Equations Algebraically (Pages 45–46)

What you should learn
How to solve polynomial equations

List four methods for solving quadratic equations:

To solve a quadratic equation by factoring, . . .

Example 6: Solve $x^2 - 12x = -27$ by factoring.

Using the Quadratic Formula to solve the quadratic equation written in general form as $ax^2 + bx + c = 0$ gives the solutions:

Example 7: For the quadratic equation $3x - 16 = -2x^2$, find the values of *a*, *b*, and *c* to be substituted into the Quadratic Formula. Then find the solutions of the equation. Round to two decimal places.

Example 8: Describe a strategy for solving the polynomial equation $x^3 + 2x^2 - x = 2$. Then find the solutions.

VI. Other Types of Equations (Pages 47–49)

What you should learn How to solve equations involving radicals, fractions, or absolute values

An equation involving a radical expression can often be cleared of radicals by . . .

When using this procedure, remember to check for

, which do not satisfy the original equation.

Example 9: Describe a strategy for solving the following equation involving a radical expression: $\sqrt{8-x} - 15 = 0$

To solve an equation involving fractions, . . .

Example 10: Solve: $\frac{2}{x} - 1 = \frac{1}{x+1}$

To solve an equation involving an absolute value, . . .

Example 11: Write the two equations that must be solved to solve the absolute value equation $\left|3x^2 + 2x\right| - 5 = 0$.

Homework Assignment Page(s) Exercises

Course Number

Instructor

Date

Section P.5 Solving Inequalities Algebraically and Graphically

Objective: In this lesson you learned how to solve linear inequalities, inequalities involving absolute values, polynomial inequalities, and rational inequalities.

Important Vocabulary Define each term or concept.

Solutions of an inequality

Graph of an inequality

Linear inequality

Double inequality

Critical numbers

Test intervals

I. Properties of Inequalities (Pages 54–55)

What you should learn
How to recognize properties of inequalities

Solving an inequality in the variable x means . . .

Numbers that are solutions of an inequality are said to

_____ the inequality.

To solve a linear inequality in one variable, use the

_____ to isolate the variable.

When both sides of an inequality are multiplied or divided by a negative number, . . .

Two inequalities that have the same solution set are

_____.

Complete the list of Properties of Inequalities given below.

1) Transitive Property: $a < b$ and $b < c \rightarrow$

2) Addition of Inequalities: $a < b$ and $c < d \rightarrow$

3) Addition of a Constant c: $a < b \rightarrow$

4) Multiplication by a Constant c:

For $c > 0$, $a < b \rightarrow$

For $c < 0$, $a < b \rightarrow$

II. Solving a Linear Inequality (Pages 55–56)

> ***What you should learn***
> How to use properties of inequalities to solve linear inequalities

Describe the steps that would be necessary to solve the linear inequality $7x - 2 < 9x + 8$.

To use a graphing utility to solve the linear inequality $7x - 2 < 9x + 8$, . . .

The two inequalities $-10 < 3x$ and $14 \geq 3x$ can be rewritten as the double inequality .

III. Inequalities Involving Absolute Value (Page 57)

> ***What you should learn***
> How to solve inequalities involving absolute values

Let x be a variable or an algebraic expression and let a be a real number such that $a \geq 0$. The solutions of $|x| < a$ are all values of x that . The solutions of $|x| > a$ are all values of x that .

Example 1: Solve the inequality: $|x + 11| - 4 \leq 0$

The symbol $\cup$ is called a symbol and is used to denote .

Example 2: Write the following solution set using interval notation: $x > 8$ or $x < 2$

IV. Polynomial Inequalities (Pages 58–60)

What you should learn
How to solve polynomial inequalities

Where can polynomials change signs?

Between two consecutive zeros, a polynomial must be . . .

When the real zeros of a polynomial are put in order, they divide the real number line into . . .

These zeros are the of the inequality, and the resulting open intervals are the

Complete the following steps for determining the intervals on which the values of a polynomial are entirely negative or entirely positive:

1)

2)

3)

To approximate the solution of the polynomial inequality $3x^2 + 2x - 5 < 0$ from a graph, . . .

If a polynomial inequality is not given in general form, you should begin the solution process by . . .

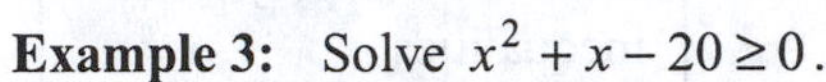

Example 3: Solve $x^2 + x - 20 \geq 0$.

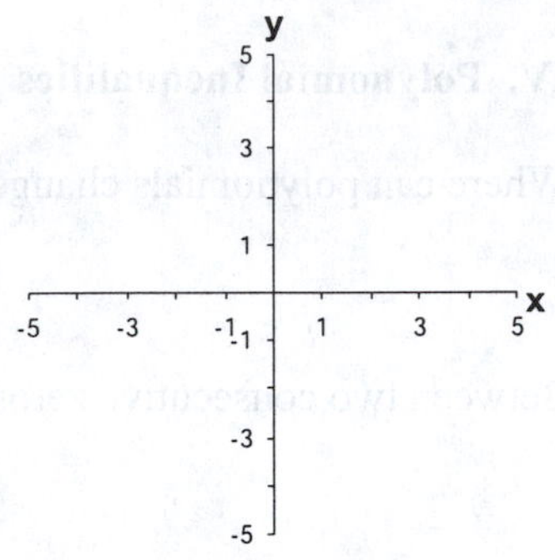

Example 4: Use a graph to solve the polynomial inequality $-x^2 - 6x - 9 > 0$.

V. Rational Inequalities (Page 61)

What you should learn
How to solve rational inequalities

To extend the concepts of critical numbers and test intervals to rational inequalities, use the fact that the value of a rational expression can change sign only at its   and its   . These two types of numbers make up the   of a rational inequality.

To solve a rational inequality, . . .

Example 5: Solve $\dfrac{3x+15}{x-2} \leq 0$.

Homework Assignment

Page(s)

Exercises

Chapter 1 Functions and Their Graphs

Course Number
Instructor
Date

Section 1.1 Functions

Objective: In this lesson you learned how to evaluate functions and find their domains.

Important Vocabulary	Define each term or concept.
Function	
Domain	
Range	
Independent variable	
Dependent variable	

I. Introduction to Functions (Pages 74–76)

What you should learn
How to decide whether relations between two variables are functions

A rule of correspondence that pairs items from one set with items from a different set is a

In functions that can be represented by ordered pairs, the first coordinate in each ordered pair is the and the second coordinate is the

Some characteristics of functions are . . .

1)

2)

3)

To decide whether a relation is a function, . . .

If any input value of a relation is matched with two or more output values, . . .

Some common ways to represent functions are . . .

1)

2)

3)

4)

Example 1: Decide whether the table represents y as a function of x.

x	-3	-1	0	2	4
y	5	-12	5	3	14

II. Function Notation (Pages 76–77)

> ***What you should learn***
> How to use function notation and evaluate functions

The symbol is **function notation** for the value of f at x or f of x, used to describe y as a function of x. In this case, is the name of the function and is the output value of the function at the input value x.

Example 2: If $f(w) = 4w^3 - 5w^2 - 7w + 13$, describe how to find $f(-2)$.

A piecewise-defined function is . . .

III. The Domain of a Function (Page 78)

> ***What you should learn***
> How to find the domains of functions

If x is in the domain of f, then f is said to be at x.

If x is not in the domain of f, then f is said to be at x.

The **implied domain** of a function defined by an algebraic expression is . . .

List and describe three common approaches to solving a problem.

1)

2)

3)

III. Applications of Graphs of Equations (Pages 73–74)

What you should learn
How to use graphs of equations in real-life problems

Describe a real-life situation in which a graphical solution approach would be helpful.

Example 4: Suppose a toy company estimates that its top-selling toy sells 240 units per minute, on average, nationally during the holiday shopping season, or according to the equation $S = 240m$, where S is the number of units sold and m is the number of minutes. Explain how a graphing utility could be used to find how long it takes during the holiday shopping season to sell 82,800 units.

Additional notes

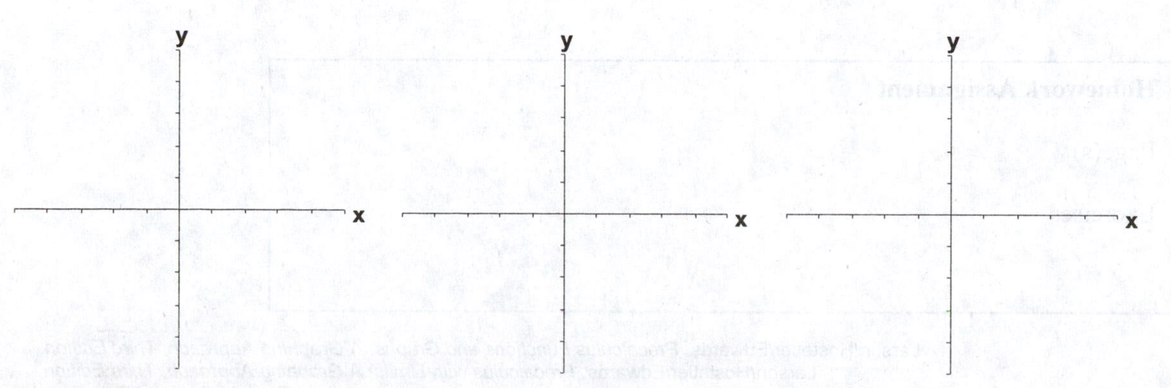

Additional notes

y x y x y x

Homework Assignment

Page(s)

Exercises

Course Number

Instructor

Date

Section 1.2 Graphs of Functions

Objective: In this lesson you learned how to analyze the graphs of functions.

Important Vocabulary	Define each term or concept.
Graph of a function	
Greatest integer function	
Step function	
Even function	
Odd function	

I. The Graph of a Function (Pages 88–89)

What you should learn
How to find the domains and ranges of functions and how to use the Vertical Line Test for functions

Explain the use of open or closed dots in the graphs of functions.

To find the domain of a function from its graph, . . .

To find the range of a function from its graph, . . .

The **Vertical Line Test** for functions states . . .

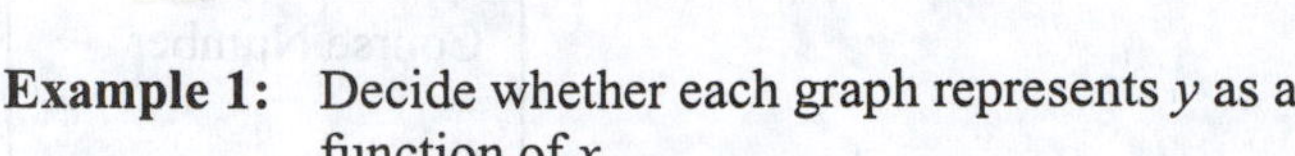

Example 1: Decide whether each graph represents y as a function of x.

(a)

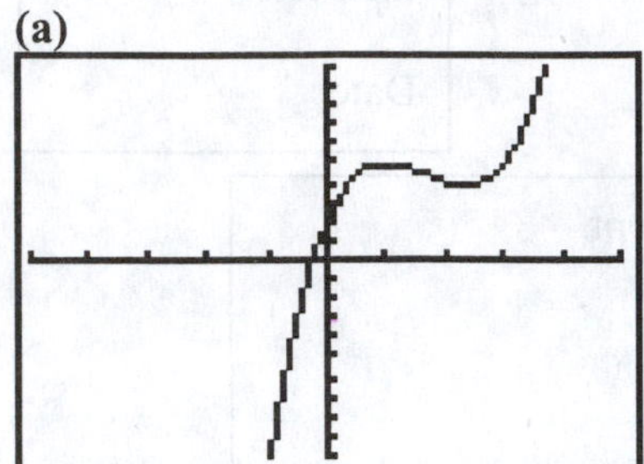

(b)

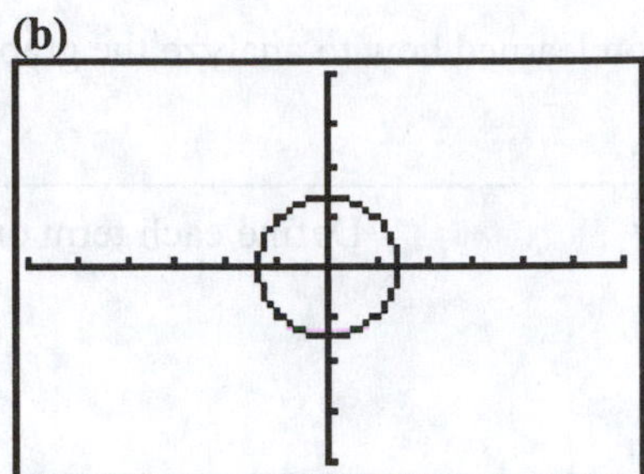

II. Increasing and Decreasing Functions (Page 90)

> ***What you should learn***
> How to determine intervals on which functions are increasing or decreasing

A function f is **increasing** on an interval if, for any x_1 and x_2 in the interval, . . .

A function f is **decreasing** on an interval if, for any x_1 and x_2 in the interval, . . .

A function f is **constant** on an interval if, for any x_1 and x_2 in the interval,. . .

Given a graph of a function, to find an interval on which the function is increasing . . .

Given a graph of a function, to find an interval on which the function is decreasing . . .

Given a graph of a function, to find an interval on which the function is constant . . .

III. Relative Minimum and Maximum Values (Pages 91–92)

> ***What you should learn***
> How to determine relative maximum and relative minimum values of functions

A function value $f(a)$ is called a **relative minimum** of f if . . .

A function value $f(a)$ is called a **relative maximum** of f if . . .

The point at which a function changes from increasing to decreasing is a relative ________. The point at which a function changes from decreasing to increasing is a relative ________.

To approximate the relative minimum or maximum of a function using a graphing utility, . . .

Example 2: Suppose a function C represents the annual number of cases (in millions) of chicken pox reported for the year x in the United States from 1960 through 2000. Interpret the meaning of the function's minimum at (1998, 3).

IV. Graphing Step Functions and Piecewise-Defined Functions (Page 93)

What you should learn
How to identify and graph step functions and other piecewise-defined functions

Describe the graph of the greatest integer function.

Example 3: Let $f(x) = [\![x]\!]$, the greatest integer function. Find $f(3.74)$.

To sketch the graph of a piecewise-defined function, . . .

V. Even and Odd Functions (Pages 94–95)

What you should learn
How to identify even and odd functions

A graph is symmetric with respect to the y-axis if, whenever (x, y) is on the graph, ______________ is also on the graph. A graph is symmetric with respect to the x-axis if, whenever (x, y) is on the graph, ______________ is also on the graph. A graph is symmetric with respect to the origin if, whenever (x, y) is on the graph, ______________ is also on the graph.

A function whose graph is symmetric with respect to the y-axis is a(n) ______________ function. A function whose graph is symmetric with respect to the origin is a(n) ______________ function. The graph of a (nonzero) function cannot be symmetric with respect to the ______________.

Additional notes

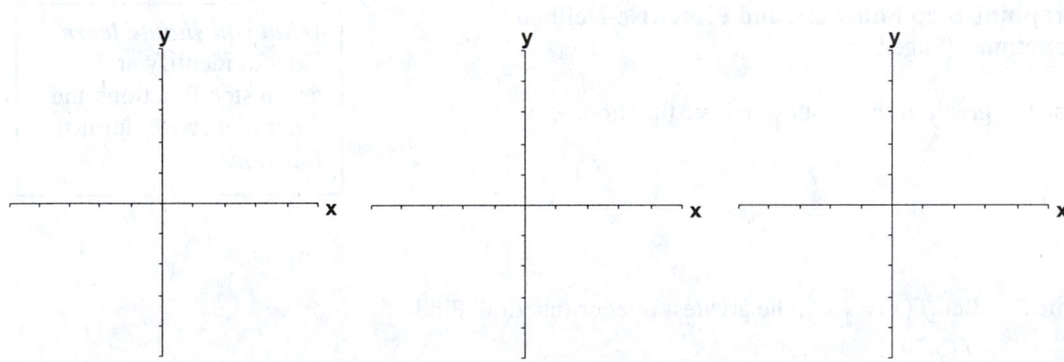

Homework Assignment

Page(s)

Exercises

Course Number

Instructor

Date

Section 1.3 Shifting, Reflecting, and Stretching Graphs

Objective: In this lesson you learned how to identify and graph shifts, reflections, and nonrigid transformations of functions.

Important Vocabulary Define each term or concept.

Vertical shift

Horizontal shift

Rigid transformations

Nonrigid transformations

I. Summary of Graphs of Common Functions (Page 100)

What you should learn
How to recognize graphs of common functions

Sketch an example of each of the six most commonly used functions in algebra.

Constant Function Identity Function

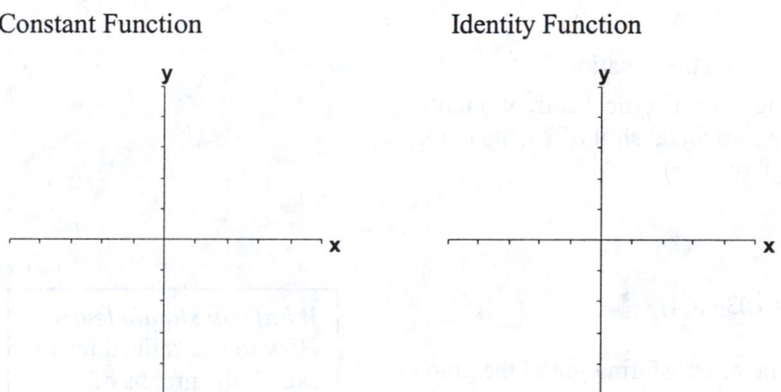

Absolute Value Function Square Root Function

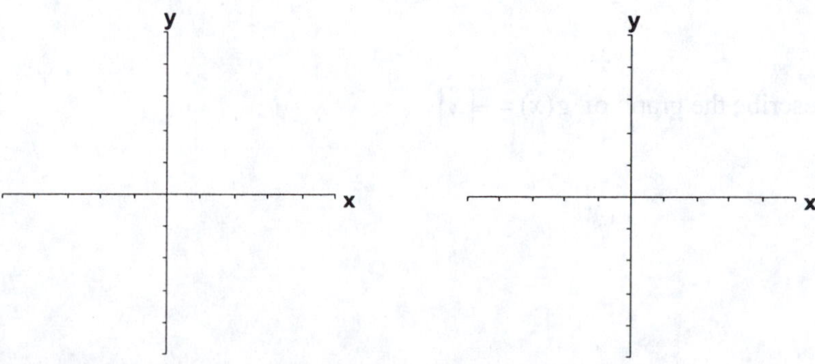

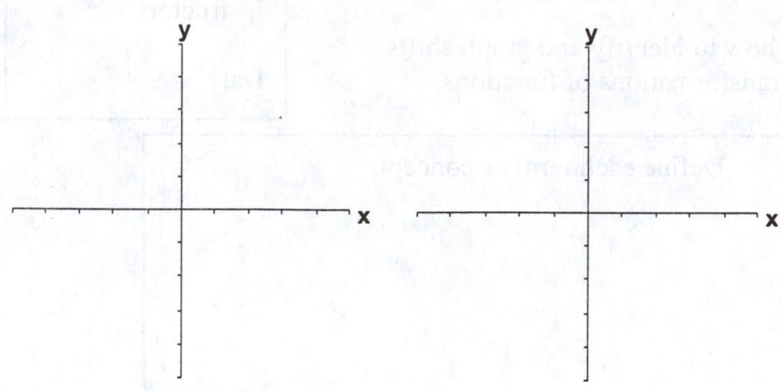

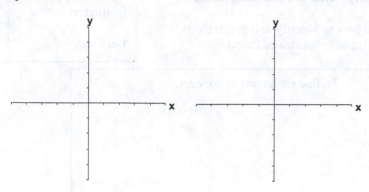

II. Vertical and Horizontal Shifts (Pages 101–102)

> ***What you should learn***
> How to use vertical and horizontal shifts to sketch the graphs of functions

Let c be a positive real number. Complete the following representations of shifts in the graph of $y = f(x)$:

1) Vertical shift c units upward:

2) Vertical shift c units downward:

3) Horizontal shift c units to the right:

4) Horizontal shift c units to the left:

Example 1: Let $f(x) = |x|$. Write the equation for the function resulting from a vertical shift of 3 units downward and a horizontal shift of 2 units to the right of the graph of $f(x)$.

III. Reflecting Graphs (Pages 103–104)

> ***What you should learn***
> How to use reflections to sketch the graphs of functions

A **reflection** in the x-axis is a type of transformation of the graph of $y = f(x)$ represented by $h(x) =$. A **reflection** in the y-axis is a type of transformation of the graph of $y = f(x)$ represented by $h(x) =$

Example 2: Let $f(x) = |x|$. Describe the graph of $g(x) = -|x|$ in terms of f.

IV. Nonrigid Transformations (Page 105)

> ***What you should learn***
> How to use nonrigid transformations to sketch graphs of functions

Name three types of rigid transformations:

1)

2)

3)

Rigid transformations change only the of the graph in the *xy*-plane.

Name two types of nonrigid transformations:

1)

2)

A nonrigid transformation $y = cf(x)$ of the graph of $y = f(x)$ is a if $c > 1$ or a if $0 < c < 1$.

Additional notes

y

x

y

x

y

x

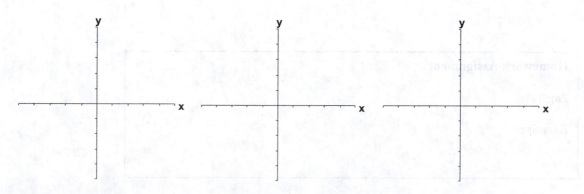

Additional notes

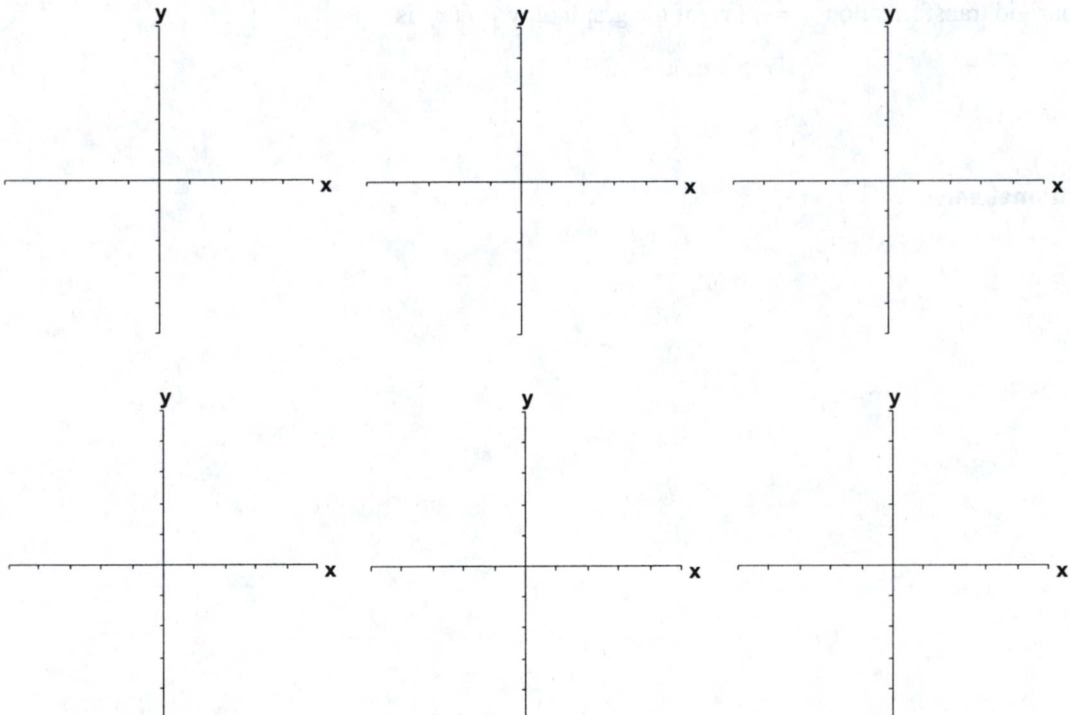

Homework Assignment

Page(s)

Exercises

Course Number
Instructor
Date

Section 1.4 Combinations of Functions

Objective: In this lesson you learned how to find arithmetic combinations and compositions of functions.

Important Vocabulary	Define each term or concept.
Arithmetic combination	
Composition	

I. Arithmetic Combinations of Functions (Pages 109–111)

> ***What you should learn***
> How to add, subtract, multiply, and divide functions

Just as two real numbers can be combined with arithmetic operations, two functions can be combined by the operations of

to create new functions.

The domain of an arithmetic combination of functions *f* and *g* consists of . . .

Let *f* and *g* be two functions with overlapping domains. Complete the following arithmetic combinations of *f* and *g* for all *x* common to both domains:

1) Sum: $(f+g)(x) =$

2) Difference: $(f-g)(x) =$

3) Product: $(fg)(x) =$

4) Quotient: $\left(\dfrac{f}{g}\right)(x) =$

To use a graphing utility to graph the sum of two functions, . . .

Example 1: Let $f(x) = 7x - 5$ and $g(x) = 3 - 2x$. Find $(f-g)(4)$.

II. Compositions of Functions (Pages 111–114)

What you should learn
How to find compositions of one function with another function

For two functions f and g, to find $(f \circ g)(x)$, . . .

For the composition of the function f with g, the domain of $f \circ g$ is . . .

Example 2: Let $f(x) = 3x + 4$ and let $g(x) = 2x^2 - 1$. Find (a) $(f \circ g)(x)$ and (b) $(g \circ f)(x)$.

III. Applications of Combinations of Functions (Page 115)

What you should learn
How to use combinations of functions to model and solve real-life problems

The function $f(x) = 0.06x$ represents the sales tax owed on a purchase with a price tag of x dollars and the function $g(x) = 0.75x$ represents the sale price of an item with a price tag of x dollars during a 25% off sale. Using one of the combinations of functions discussed in this section, write the function that represents the sales tax owed on an item with a price tag of x dollars during a 25% off sale.

Additional notes

Homework Assignment

Page(s)

Exercises

Course Number

Instructor

Date

Section 1.5 Inverse Functions

Objective: In this lesson you learned how to find inverses of functions graphically and algebraically.

Important Vocabulary Define each term or concept.

Inverse function

One-to-one

Horizontal Line Test

I. The Inverse of a Function (Pages 120–122)

What you should learn
How to find inverse functions informally and verify that two functions are inverses of each other

For a function *f* that is defined by a set of ordered pairs, to form the inverse function of *f*, . . .

For a function *f* and its inverse f^{-1}, the domain of *f* is equal to

, and the range of *f* is equal to

To verify that two functions, *f* and *g*, are inverses of each other,

. . .

Example 1: Verify that the functions $f(x) = 2x - 3$ and $g(x) = \dfrac{x+3}{2}$ are inverses of each other.

II. The Graph of an Inverse Function (Page 123)

What you should learn
How to verify graphically and numerically that two functions are inverses of each other

If the point (*a*, *b*) lies on the graph of *f*, then the point (,) lies on the graph of f^{-1} and vice versa. The graph of f^{-1} is a reflection of the graph of *f* in the line

III. The Existence of an Inverse Function (Page 124)

What you should learn
How to use graphs of functions to decide whether functions have inverses

A function *f* has an inverse f^{-1} if and only if . . .

If a function is one-to-one, that means . . .

To tell whether a function is one-to-one from its graph, . . .

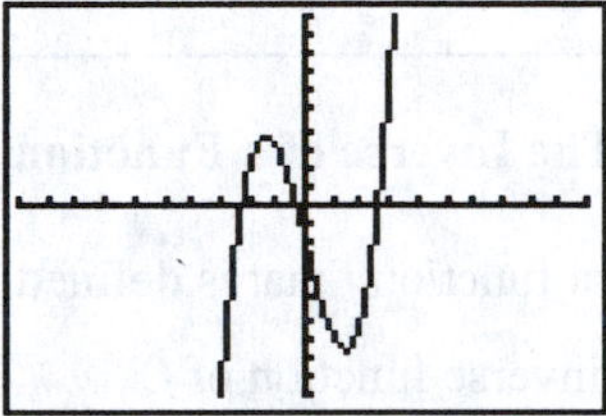

Example 2: Does the graph of the function at the right have an inverse function? Explain.

IV. Finding Inverse Functions Algebraically (Pages 125–126)

What you should learn
How to find inverse functions algebraically

To find the inverse of a function *f* algebraically, . . .

1)

2)

3)

4)

5)

Example 3: Find the inverse (if it exists) of $f(x) = 4x - 5$.

Homework Assignment

Page(s)

Exercises

Chapter 2 Polynomial and Rational Functions

Course Number

Instructor

Date

Section 2.1 Quadratic Functions

Objective: In this lesson you learned how to sketch and analyze graphs of quadratic functions.

Important Vocabulary Define each term or concept.

Constant function

Linear function

Quadratic function

Axis of symmetry

Vertex

I. The Graph of a Quadratic Function (Pages 136–138)

What you should learn
How to analyze graphs of quadratic functions

Let n be a nonnegative integer and let $a_n, a_{n-1}, \ldots, a_2, a_1, a_0$ be real numbers with $a_n \neq 0$. A **polynomial function of x with degree n** is . . .

A quadratic function is a polynomial function of degree. The graph of a quadratic function is a special "U"-shaped curve called a

If the leading coefficient of a quadratic function is positive, the graph of the function opens and the vertex of the parabola is the point on the graph. If the leading coefficient of a quadratic function is negative, the graph of the function opens and the vertex of the parabola is the point on the graph.

II. The Standard Form of a Quadratic Function (Pages 139–140)

What you should learn
How to write quadratic functions in standard form and use the results to sketch graphs of functions

The **standard form of a quadratic function** is

______________________________________.

For a quadratic function in standard form, the axis of the associated parabola is   and the vertex is

To write a quadratic function in standard form, . . .

To find the x-intercepts of the graph of $f(x) = ax^2 + bx + c$, . . .

Example 1: Sketch the graph of $f(x) = x^2 + 2x - 8$ and identify the vertex, axis, and x-intercepts of the parabola.

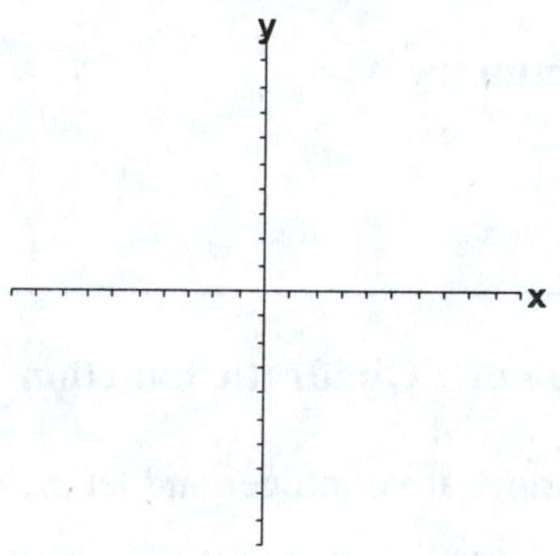

III. Applications (Pages 141–142)

What you should learn
How to use quadratic functions to model and solve real-life problems

For a quadratic function in the form $f(x) = ax^2 + bx + c$, when $a > 0$, f has a minimum that occurs at

When $a < 0$, f has a maximum that occurs at

To find the minimum or maximum value,

Example 2: Find the minimum value of the function $f(x) = 3x^2 - 11x + 16$. At what value of x does this minimum occur?

Homework Assignment

Page(s)

Exercises

Course Number
Instructor
Date

Section 2.2 Polynomial Functions of Higher Degree

Objective: In this lesson you learned how to sketch and analyze graphs of polynomial functions.

Important Vocabulary Define each term or concept.

Continuous

Extrema

Relative minimum

Relative maximum

Repeated zero

Multiplicity

Intermediate Value Theorem

I. Graphs of Polynomial Functions (Pages 147–148)

What you should learn
How to use transformations to sketch graphs of polynomial functions

Name two basic features of the graphs of polynomial functions.

1)
2)

Will the graph of $g(x) = x^7$ look more like the graph of $f(x) = x^2$ or the graph of $f(x) = x^3$? Explain.

II. The Leading Coefficient Test (Pages 149–150)

What you should learn
How to use the Leading Coefficient Test to determine the end behavior of graphs of polynomial functions

State the **Leading Coefficient Test.**

Example 1: Describe the left and right behavior of the graph of $f(x) = 1 - 3x^2 - 4x^6$.

III. Zeros of Polynomial Functions (Pages 150–154)

What you should learn
How to find and use zeros of polynomial functions to sketch their graphs

Let *f* be a polynomial function of degree *n*. The function *f* has at most real zeros. The graph of *f* has at most relative extrema.

Let *f* be a polynomial function and let *a* be a real number. List four equivalent statements about the real zeros of *f*.

1)

2)

3)

4)

If a polynomial function *f* has a repeated zero $x = 3$ with multiplicity 4, the graph of *f* the *x*-axis at $x =$. If *f* has a repeated zero $x = 4$ with multiplicity 3, the graph of *f* the *x*-axis at $x =$.

y

x

Example 2: Sketch the graph of $f(x) = \frac{1}{4}x^4 - 2x^2 + 3$.

IV. The Intermediate Value Theorem (Pages 154–155)

What you should learn
How to use the Intermediate Value Theorem to help locate zeros of polynomial functions

Interpret the meaning of the Intermediate Value Theorem.

Describe how the Intermediate Value Theorem can help in locating the real zeros of a polynomial function *f*.

Homework Assignment

Page(s)

Exercises

Course Number

Instructor

Date

Section 2.3 Real Zeros of Polynomial Functions

Objective: In this lesson you learned how to use long division and synthetic division to divide polynomials by other polynomials and how to find the rational and real zeros of polynomial functions.

Important Vocabulary Define each term or concept.

Long division of polynomials

Division Algorithm

Synthetic division

Remainder Theorem

Factor Theorem

Upper bound

Lower bound

I. Long Division of Polynomials (Pages 160–162)

What you should learn
How to use long division to divide polynomials by other polynomials

When dividing a polynomial $f(x)$ by another polynomial $d(x)$, if the remainder $r(x) = 0$, $d(x)$ into $f(x)$.

The rational expression $f(x)/d(x)$ is improper if . . .

The rational expression $r(x)/d(x)$ is proper if . . .

The result of a division problem can be checked by . . .

Example 1: Divide $3x^3 + 4x - 2$ by $x^2 + 2x + 1$.

Larson/Hostetler/Edwards *Precalculus Functions and Graphs: A Graphing Approach, Third Edition*
Larson/Hostetler/Edwards *Precalculus with Limits: A Graphing Approach, Third Edition*
Student Success Organizer

II. Synthetic Division (Page 163)

What you should learn
How to use synthetic division to divide polynomials by binomials of the form $(x - k)$

Can synthetic division be used to divide a polynomial by $x^2 - 5$? Explain.

Can synthetic division be used to divide a polynomial by $x + 4$? Explain.

Example 2: Fill in the following synthetic division array to divide $2x^4 + 5x^2 - 3$ by $x - 5$. Then carry out the synthetic division and indicate which entry represents the remainder.

III. The Remainder and Factor Theorems (Pages 164–165)

What you should learn
How to use the Remainder Theorem and the Factor Theorem

To use the Remainder Theorem to evaluate a polynomial function $f(x)$ at $x = k$, . . .

Example 3: Use the Remainder Theorem to evaluate the function $f(x) = 2x^4 + 5x^2 - 3$ at $x = 5$.

To use the Factor Theorem to show that $(x - k)$ is a factor of a polynomial function $f(x)$, . . .

List three facts about the remainder r, obtained in the synthetic division of $f(x)$ by $x - k$:

1)

2)

3)

IV. The Rational Zero Test (Pages 166–168)

What you should learn
How to use the Rational Zero Test to determine possible rational zeros of polynomial functions

Describe the purpose of the Rational Zero Test.

State the **Rational Zero Test.**

To use the Rational Zero Test, . . .

Example 4: List the possible rational zeros of the polynomial function $f(x) = 3x^5 + x^4 + 4x^3 - 2x^2 + 8x - 5$.

Some strategies that can be used to shorten the search for actual zeros among a list of possible rational zeros include . . .

V. Bounds for Real Zeros of Polynomial Functions (Pages 168–169)

What you should learn
How to determine upper and lower bounds for zeros of polynomial functions

State the Upper and Lower Bound Rules.

Explain how the Upper and Lower Bound Rules can be useful in the search for the real zeros of a polynomial function.

Additional notes

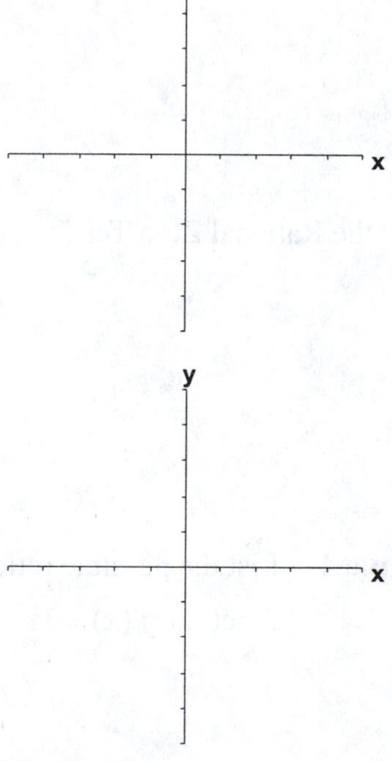

Homework Assignment

Page(s)

Exercises

Course Number

Instructor

Date

Section 2.4 Complex Numbers

Objective: In this lesson you learned how to perform operations with complex numbers and plot complex numbers in the complex plane.

Important Vocabulary Define each term or concept.

Complex number

Complex conjugates

Complex plane

Imaginary axis

Real axis

I. The Imaginary Unit *i* (Page 174)

What you should learn
How to use the imaginary unit i to write complex numbers

Mathematicians created an expanded system of numbers using the **imaginary unit *i*,** defined as $i =$, because . . .

By definition, $i^2 =$

For the complex number $a + bi$, if $b = 0$, the number $a + bi = a$ is a(n) . If $b \neq 0$, the number $a + bi$ is a(n) . If $a = 0$, the number $a + bi = bi$ is a(n)

The set of complex numbers consists of the set of and the set of

Two complex numbers $a + bi$ and $c + di$, written in standard form, are equal to each other if . . .

II. Operations with Complex Numbers (Pages 175–176)

> ***What you should learn***
> How to add, subtract, and multiply complex numbers

To add two complex numbers, . . .

To subtract two complex numbers, . . .

The additive identity in the complex number system is

The additive inverse of the complex number $a + bi$ is

Example 1: Perform the operations:
$(5 - 6i) - (3 - 2i) + 4i$

To multiply two complex numbers $a + bi$ and $c + di$, . . .

Example 2: Multiply: $(5 - 6i)(3 - 2i)$

III. Complex Conjugates and Division (Page 177)

> ***What you should learn***
> How to use complex conjugates to divide complex numbers

The product of a pair of complex conjugates is a(n)

number.

To find the quotient of the complex numbers $a + bi$ and $c + di$, where c and d are not both zero, . . .

Example 3: Divide $(1 + i)$ by $(2 - i)$. Write the result in standard form.

IV. Applications of Complex Numbers (Pages 178–179)

What you should learn
How to plot complex numbers in the complex plane

On the complex plane shown below, (a) label the real axis, (b) label the imaginary axis, and (c) plot and label the complex numbers $-2-3i$ and $4+i$.

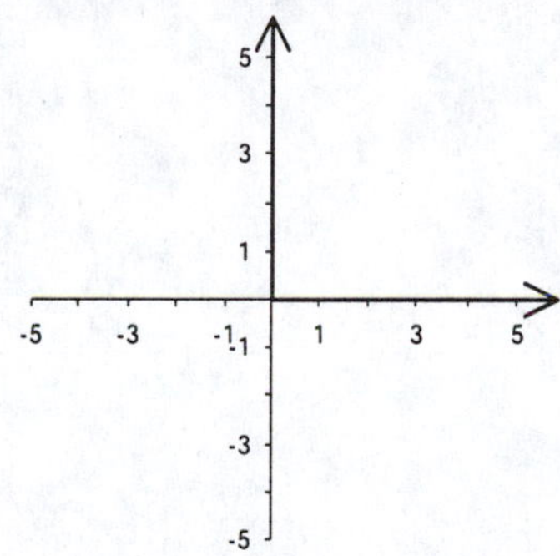

Let c represent a complex number. Describe how to tell whether or not c belongs to the Mandelbrot Set.

Describe how the Mandelbrot Set could be graphed.

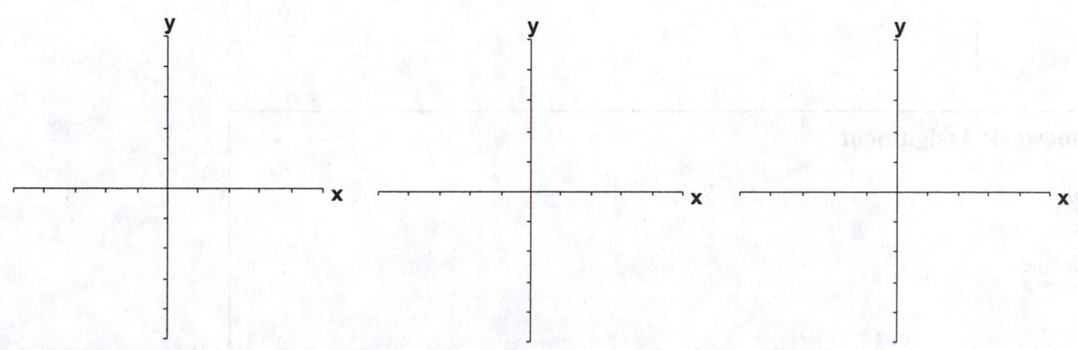

Additional notes

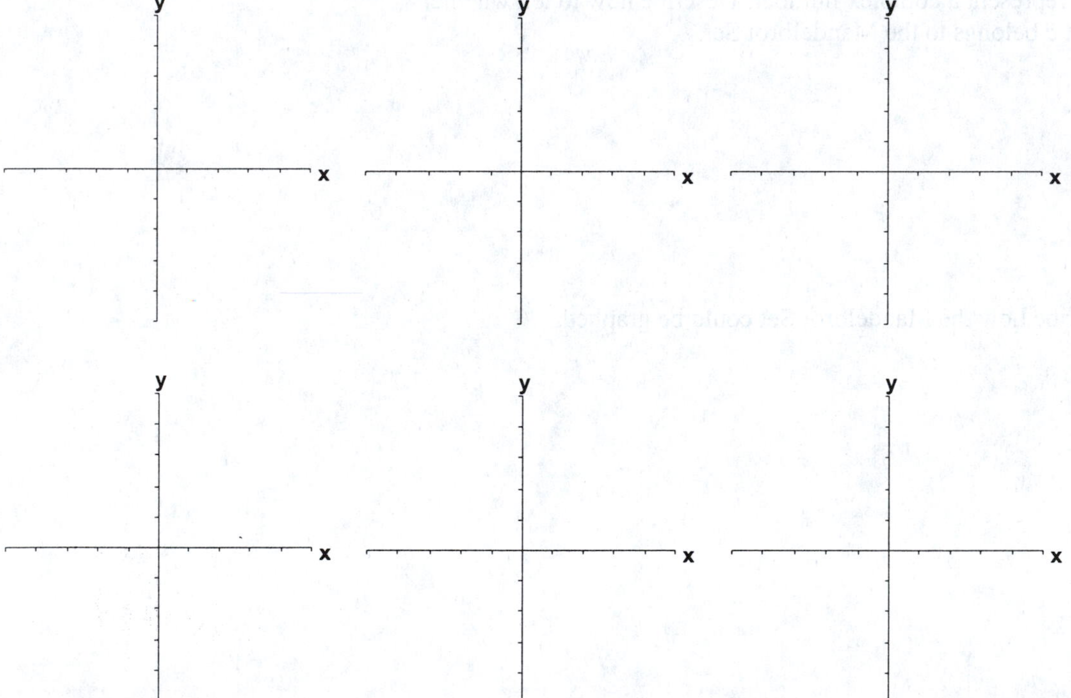

Homework Assignment

Page(s)

Exercises

Course Number

Instructor

Date

Section 2.5 The Fundamental Theorem of Algebra

Objective: In this lesson you learned how to determine the number of zeros of polynomial functions and find them.

Important Vocabulary Define each term or concept.

Fundamental Theorem of Algebra

Linear Factorization Theorem

I. The Fundamental Theorem of Algebra (Pages 182–183)

What you should learn
How to determine the number of zeros of polynomial functions and how to find all zeros of polynomial functions, including complex zeros

In the complex number system, every *n*th-degree polynomial function has zeros.

Example 1: How many zeros does the polynomial function $f(x) = 5 - 2x^2 + x^3 - 12x^5$ have?

An *n*th-degree polynomial can be factored into linear factors.

Example 2: List all of the zeros of the polynomial function $f(x) = x^3 - 2x^2 + 36x - 72$.

II. Conjugate Pairs (Page 184)

What you should learn
How to find conjugate pairs of complex zeros

Let *f*(*x*) be a polynomial function that has real coefficients. If $a + bi$, where $b \neq 0$, is a zero of the function, then we know that is also a zero of the function.

III. Factoring a Polynomial (Pages 184–186)

What you should learn
How to find zeros of polynomials by factoring

To write a polynomial of degree $n > 0$ with real coefficients as a product without complex factors, write the polynomial as . . .

A quadratic factor with no real zeros is said to be

.

Example 3: Write the polynomial $f(x) = x^4 + 5x^2 - 36$
(a) as the product of linear factors and quadratic factors that are irreducible over the reals, and
(b) in completely factored form.

Explain why a graph cannot be used to locate complex zeros.

Additional notes

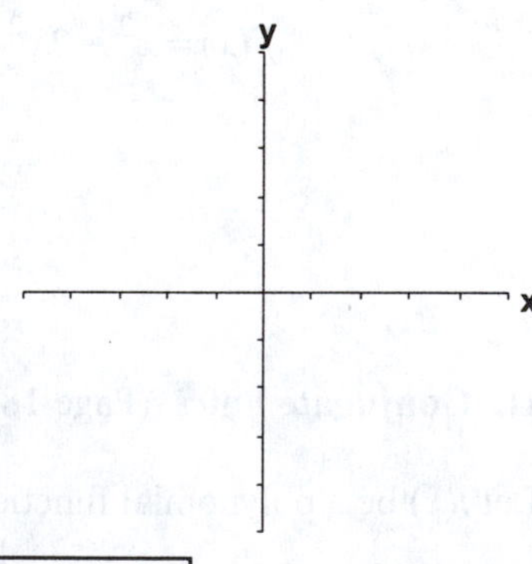

Homework Assignment

Page(s)

Exercises

Course Number

Instructor

Date

Section 2.6 Rational Functions and Asymptotes

Objective: In this lesson you learned how to determine the domain and find asymptotes of rational functions.

Important Vocabulary	Define each term or concept.
Rational function	
Vertical asymptote	
Horizontal asymptote	

I. Introduction to Rational Functions (Page 189)

What you should learn
How to find domains of rational functions

The domain of a rational function of x includes all real numbers except . . .

To find the domain of a rational function of x, . . .

Example 1: Find the domain of the function $f(x) = \dfrac{1}{x^2 - 9}$.

II. Horizontal and Vertical Asymptotes (Pages 190–192)

What you should learn
How to find horizontal and vertical asymptotes of graphs of rational functions

The notation "$f(x) \to 5$ as $x \to \infty$" means . . .

Describe the end behavior of a rational function in relation to its horizontal asymptote.

Larson/Hostetler/Edwards *Precalculus Functions and Graphs: A Graphing Approach, Third Edition*
Larson/Hostetler/Edwards *Precalculus with Limits: A Graphing Approach, Third Edition*
Student Success Organizer

Let f be the rational function given by

$$f(x) = \frac{N(x)}{D(x)} = \frac{a_n x^n + a_{n-1}x^{n-1} + \cdots + a_1 x + a_0}{b_m x^m + b_{m-1}x^{m-1} + \cdots + b_1 x + b_0}$$

where $N(x)$ and $D(x)$ have no common factors.

1) The graph of f has vertical asymptotes at

2) The graph of f has at most one horizontal asymptote determined by

.

a) If $n < m$,

b) If $n = m$, ________________________

________________.

c) If $n > m$, the graph of f has

Example 2: Find the asymptotes of the function

$$f(x) = \frac{2x - 1}{x^2 - x - 6}.$$

III. Applications of Rational Functions (Pages 193–194)

What you should learn
How to use rational functions to model and solve real-life problems

Give an example of asymptotic behavior that occurs in real life.

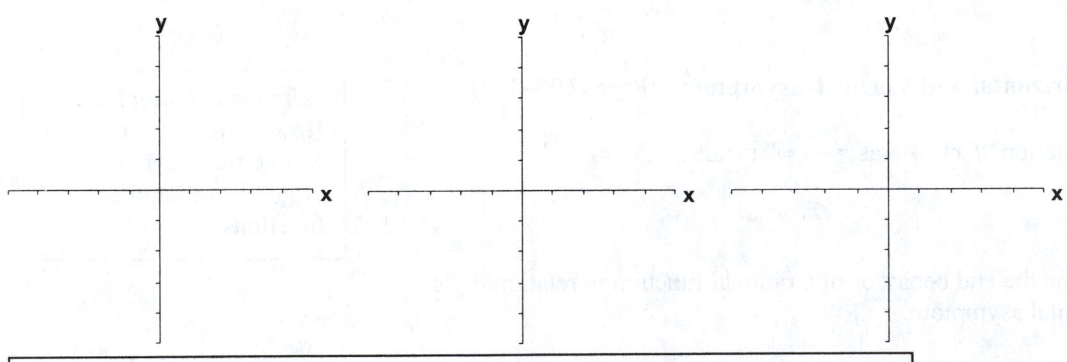

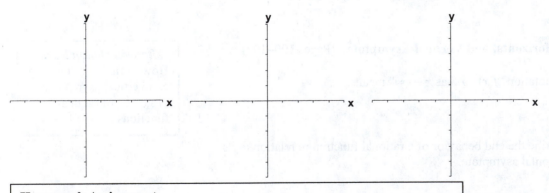

Homework Assignment

Page(s)

Exercises

Course Number

Instructor

Date

Section 2.7 Graphs of Rational Functions

Objective: In this lesson you learned how to sketch graphs of rational functions.

Important Vocabulary Define each term or concept.

Slant (or oblique) asymptote

I. The Graph of a Rational Function (Pages 199–201)

What you should learn
How to analyze and sketch graphs of rational functions

To sketch the graph of the rational function $f(x) = N(x)/D(x)$, where $N(x)$ and $D(x)$ are polynomials with no common factors, . . .

Example 1: Sketch the graph of $f(x) = \dfrac{3x}{x+4}$.

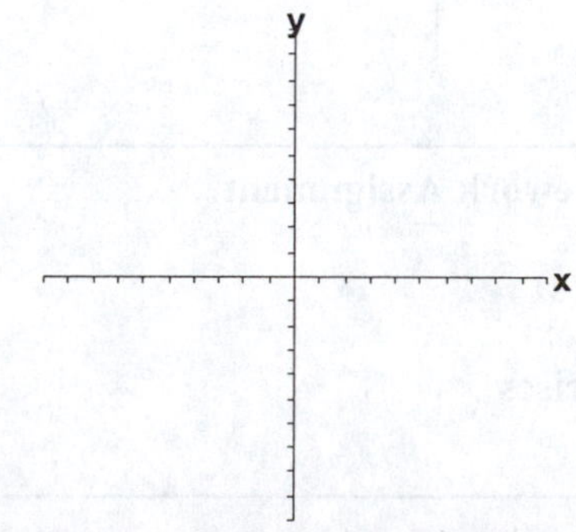

II. Slant Asymptotes (Page 202)

To find the equation of a slant asymptote, . . .

> ***What you should learn***
> How to decide whether graphs of rational functions have slant asymptotes

Example 2: Decide whether each of the following rational functions has a slant asymptote. If so, find the equation of the slant asymptote.

(a) $f(x) = \dfrac{x^3 - 1}{x^2 + 3x + 5}$ (b) $f(x) = \dfrac{3x^3 + 2}{2x - 5}$

III. Applications of Graphs of Rational Functions (Page 203)

> ***What you should learn***
> How to use rational functions to model and solve real-life problems

Describe a real-life situation in which a graph of a rational function would be helpful when solving a problem.

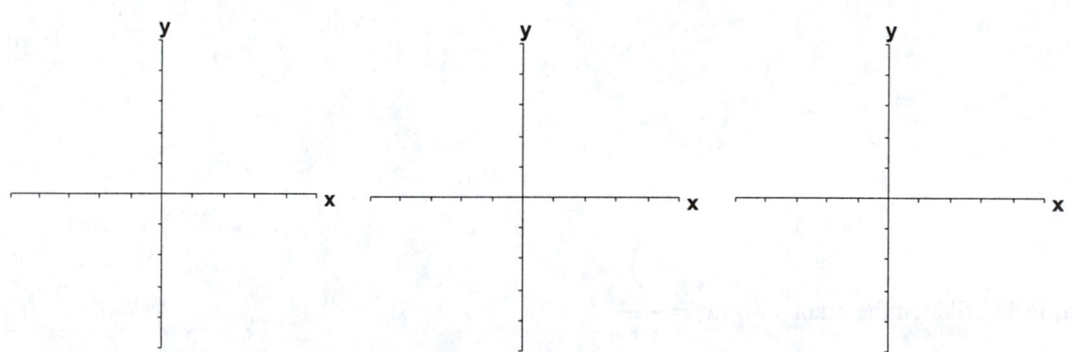

Homework Assignment

Page(s)

Exercises

Chapter 3 Exponential and Logarithmic Functions

Course Number
Instructor
Date

Section 3.1 Exponential Functions and Their Graphs

Objective: In this lesson you learned how to recognize, evaluate, and graph exponential functions.

Important Vocabulary	Define each term or concept.
Algebraic functions	
Transcendental functions	
Natural base *e*	

I. Exponential Functions (Page 216)

> ***What you should learn***
> How to recognize and evaluate exponential functions with base a

The **exponential function *f* with base *a*** is denoted by ______, where $a > 0$, $a \neq 1$, and x is any real number.

Example 1: Use a calculator to evaluate the expression $5^{3/5}$.

II. Graphs of Exponential Functions (Pages 217–219)

> ***What you should learn***
> How to graph exponential functions

For $a > 1$, is the graph of $y = a^x$ increasing or decreasing over its domain?

For $a > 1$, is the graph of $y = a^{-x}$ increasing or decreasing over its domain?

For the graph of $y = a^x$ or $y = a^{-x}$, $a > 1$, the domain is , the range is , and the intercept is . Also, both graphs have as a horizontal asymptote.

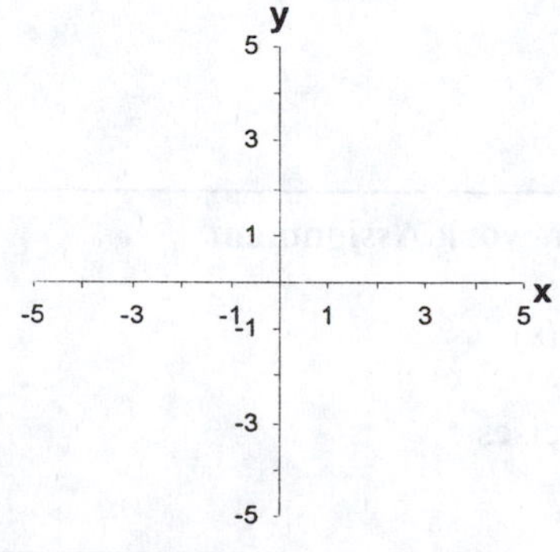

Example 2: Sketch the graph of the function $f(x) = 3^{-x}$.

III. The Natural Base e (Pages 220–221)

> ***What you should learn***
> How to recognize, evaluate, and graph exponential functions with base e

The **natural exponential function** is given by the function

.

Example 3: Use a calculator to evaluate the expression $e^{3/5}$.

For the graph of $f(x) = e^x$, the domain is ,

the range is , and the intercept is .

The number e can be approximated by the expression

for large values of x.

IV. Compound Interest and Other Applications (Pages 222–224)

> ***What you should learn***
> How to use exponential functions to model and solve real-life problems

After t years, the balance A in an account with principal P and annual interest rate r (in decimal form) is given by the formulas:

For n compoundings per year:

For continuous compounding:

Example 4: Find the amount in an account after 10 years if \$6000 is invested at an interest rate of 7%,
(a) compounded monthly.
(b) compounded continuously.

Homework Assignment

Page(s)

Exercises

Course Number

Instructor

Date

Section 3.2 Logarithmic Functions and Their Graphs

Objective: In this lesson you learned how to recognize, evaluate, and graph logarithmic functions.

Important Vocabulary Define each term or concept.

Common logarithmic function

Natural logarithmic function

I. Logarithmic Functions (Pages 229–230)

What you should learn
How to recognize and evaluate logarithmic functions with base a

The **logarithmic function with base *a*** is defined as

______________________, for $x > 0$ and $0 < a \neq 1$, if and only if $x = a^y$.

The logarithmic function with base a is the

of the exponential function $f(x) = a^x$.

The equation $x = a^y$ in exponential form is equivalent to the equation ______________________ in logarithmic form.

When evaluating logarithms, remember that a logarithm is a(n)

. This means that $\log_a x$ is the

to which a must be raised to obtain .

Example 1: Use the definition of logarithmic function to evaluate $\log_5 125$.

Example 2: Use a calculator to evaluate $\log_{10} 300$.

Complete the following properties of logarithms:

1) $\log_a 1 =$ 2) $\log_a a =$

3) $\log_a a^x =$ and $a^{\log_a x} =$

4) If $\log_a x = \log_a y$, then .

Larson/Hostetler/Edwards *Precalculus Functions and Graphs: A Graphing Approach, Third Edition*
Larson/Hostetler/Edwards *Precalculus with Limits: A Graphing Approach, Third Edition*
Student Success Organizer

Example 3: Solve the equation $\log_7 x = 1$ for *x*.

II. Graphs of Logarithmic Functions (Pages 231–232)

> ***What you should learn***
> How to graph logarithmic functions

For $a > 1$, is the graph of $y = \log_a x$ increasing or decreasing over its domain?

For the graph of $y = \log_a x$, $a > 1$, the domain is ______, the range is ______, and the intercept is ______.

Also, the graph has ______ as a vertical asymptote. The graph of $y = \log_a x$ is a reflection of the graph of $y = a^x$ about ______.

Example 4: Sketch the graph of the function $f(x) = \log_3 x$.

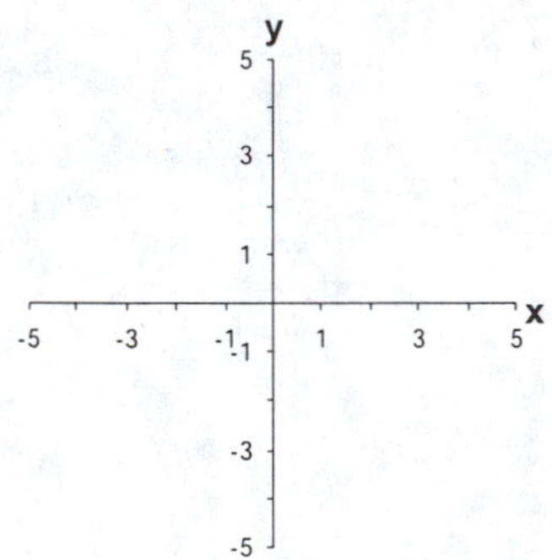

III. The Natural Logarithmic Function (Pages 233–234)

> ***What you should learn***
> How to recognize, evaluate, and graph natural logarithmic functions

Complete the following properties of natural logarithms:

1) $\ln 1 =$ 2) $\ln e =$

3) $\ln e^x =$ and $e^{\ln x} =$

4) If $\ln x = \ln y$, then ______.

Example 5: Use a calculator to evaluate $\ln 10$.

Example 6: Find the domain of the function $f(x) = \ln(x+3)$.

IV. Applications of Logarithmic Functions (Page 235)

What you should learn
How to use logarithmic functions to model and solve real-life problems

Describe a real-life situation in which logarithms are used.

Example 7: A principal P, invested at 6% interest and compounded continuously, increases to an amount K times the original principal after t years, where t is given by $t = \frac{\ln K}{0.06}$. How long will it take the original investment to double in value? To triple in value?

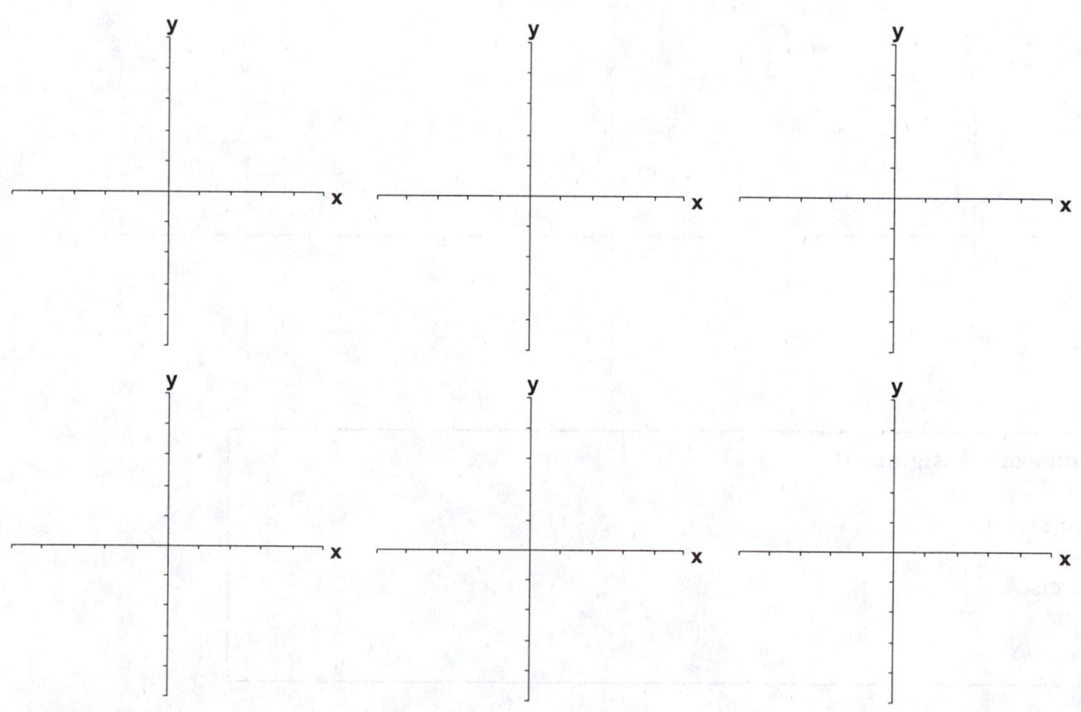

Additional notes

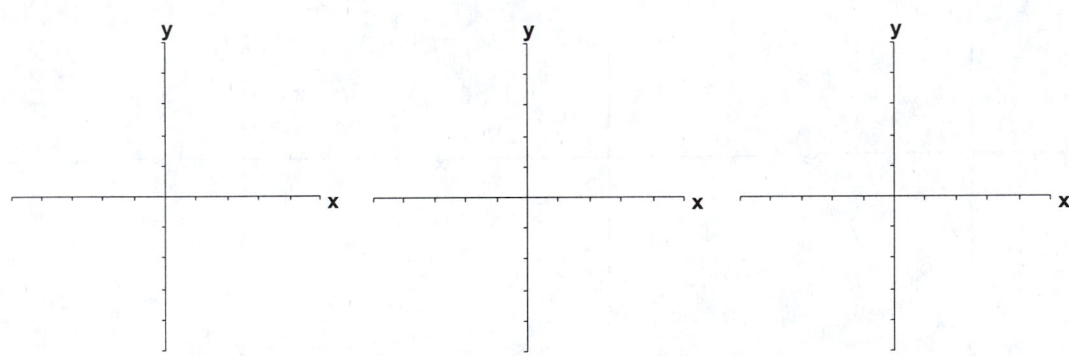

Homework Assignment

Page(s)

Exercises

Course Number
Instructor
Date

Section 3.3 Properties of Logarithms

Objective: In this lesson you learned how to rewrite logarithmic functions with different bases and how to use properties of logarithms to evaluate, rewrite, expand, or condense logarithmic expressions.

I. Change of Base (Page 240)

What you should learn
How to rewrite logarithmic functions with different bases

Let a, b, and x be positive real numbers such that $a \neq 1$ and $b \neq 1$. The **change-of-base formula** states that . . .

Explain how to use a calculator to evaluate $\log_8 20$.

II. Properties of Logarithms (Page 241)

What you should learn
How to use properties of logarithms to evaluate or rewrite logarithmic expressions

Let a be a positive number such that $a \neq 1$; let n be a real number; and let u and v be positive real numbers. Complete the following properties of logarithms:

1. $\log_a(uv) =$ ____________________

2. $\log_a \dfrac{u}{v} =$ ____________________

3. $\log_a u^n =$ ____________________

III. Rewriting Logarithmic Expressions (Page 242)

What you should learn
How to use properties of logarithms to expand or condense logarithmic expressions

To expand a logarithmic expression means to . . . .

Example 1: Expand the logarithmic expression $\ln \dfrac{xy^4}{2}$.

To condense a logarithmic expression means to . . . .

Example 2: Condense the logarithmic expression $3 \log x + 4 \log(x - 1)$.

IV. Applications of Properties of Logarithms (Page 243)

> ***What you should learn***
> How to use logarithmic functions to model and solve real-life problems

One way of finding a model for a set of nonlinear data is to take the natural log of each of the *x*-values and *y*-values of the data set. If the points are graphed and fall on a straight line, then the *x*-values and the *y*-values are related by the equation:

, where *m* is the slope of the straight line.

Example 3: Find a natural logarithmic equation for the following data that expresses *y* as a function of *x*.

x	2.718	7.389	20.086	54.598
y	7.389	54.598	403.429	2980.958

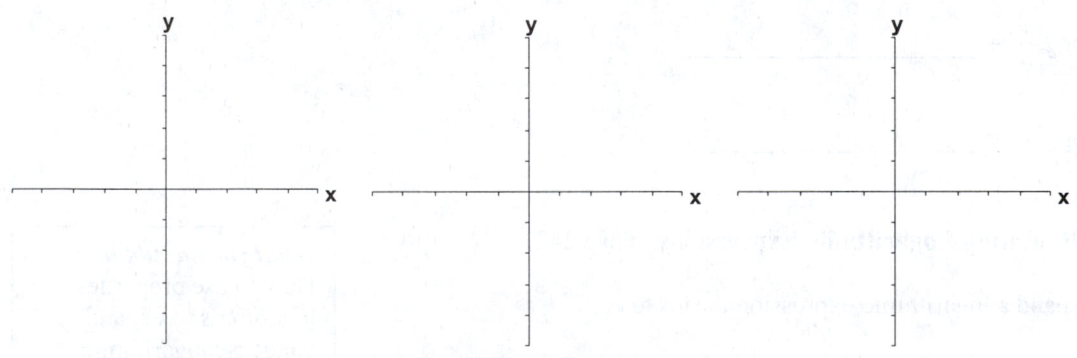

Homework Assignment

Page(s)

Exercises

Section 3.4 Solving Exponential and Logarithmic Equations

Course Number

Instructor

Date

Objective: In this lesson you learned how to solve exponential and logarithmic equations.

I. Introduction (Page 247)

What you should learn
How to solve simple exponential and logarithmic equations

State the One-to-One Property for exponential equations.

State the One-to-One Property for logarithmic equations.

State the Inverse Properties for exponential equations and for logarithmic equations.

Describe how the One-to-One Properties and the Inverse Properties can be used to solve exponential and logarithmic equations.

Example 1: (a) Solve $\log_8 x = \frac{1}{3}$ for x.
(b) Solve $5^x = 0.04$ for x.

II. Solving Exponential Equations (Pages 248–249)

What you should learn
How to solve more complicated exponential equations

Describe how to solve the exponential equation $10^x = 90$ algebraically.

Larson/Hostetler/Edwards *Precalculus Functions and Graphs: A Graphing Approach, Third Edition*
Larson/Hostetler/Edwards *Precalculus with Limits: A Graphing Approach, Third Edition*
Student Success Organizer

Example 2: Solve $e^{x-2} - 7 = 59$ for x. Round to three decimal places.

III. Solving Logarithmic Equations (Pages 250–252)

What you should learn
How to solve more complicated logarithmic equations

Describe how to solve the logarithmic equation $\log_6(4x-7) = \log_6(8-x)$ algebraically.

Example 3: Solve $4 \ln 5x = 28$ for x. Round to three decimal places.

IV. Approximating Solutions (Page 252)

What you should learn
How to approximate the solutions of exponential or logarithmic equations with a graphing utility

Describe at least two different methods that can be used to approximate the solutions of an exponential or logarithmic equation using a graphing utility.

V. Applications of Solving Exponential and Logarithmic Equations (Page 253)

What you should learn
How to use exponential and logarithmic equations to model and solve real-life problems

Example 4: Use the formula for continuous compounding, $A = Pe^{rt}$, to find how long it will take \$1500 to triple in value if it is invested at 12% interest, compounded continuously.

Homework Assignment

Page(s)

Exercises

Course Number
Instructor
Date

Section 3.5 Exponential and Logarithmic Models

Objective: In this lesson you learned how to use exponential growth models, exponential decay models, Gaussian models, logistic models, and logarithmic models to solve real-life problems and how to fit exponential and logarithmic models to sets of data.

Important Vocabulary Define each term or concept.

Bell-shaped curve

Logistic curve

Sigmoidal curve

I. Introduction (Page 258)

What you should learn
How to recognize the five most common types of models involving exponential and logarithmic functions

The **exponential growth model** is .

The **exponential decay model** is .

The **Gaussian model** is .

The **logistic growth model** is .

Logarithmic models are and

______________.

II. Exponential Growth and Decay (Pages 259–261)

What you should learn
How to use exponential growth and decay functions to model and solve real-life problems

Example 1: Suppose a population is growing according to the model $P = 800e^{0.05t}$, where t is given in years.
(a) What is the initial size of the population?
(b) How long will it take this population to double?

To estimate the age of dead organic matter, scientists use the carbon dating model , which denotes the ratio R of carbon 14 to carbon 12 present at any time t (in years).

Example 2: The ratio of carbon 14 to carbon 12 in a fossil is $R = 10^{-16}$. Find the age of the fossil.

III. Gaussian Models (Page 262)

What you should learn
How to use Gaussian functions to model and solve real-life problems

The Gaussian model is commonly used in probability and statistics to represent populations that are

On a bell-shaped curve, the average value for a population is the
of the curve.

Example 3: Draw the basic form of the graph of a Gaussian model.

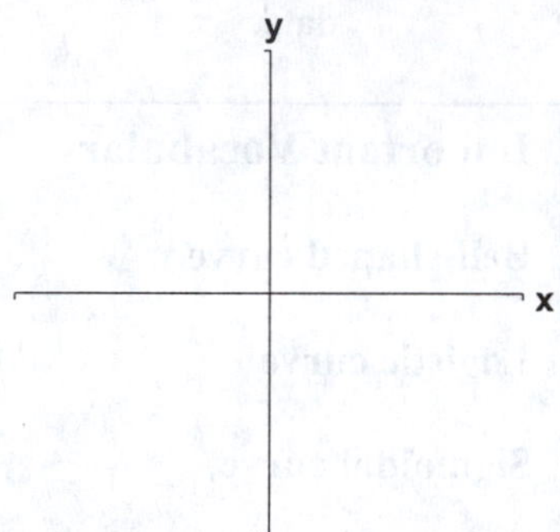

IV. Logistic Growth Models (Page 263)

What you should learn
How to use logistic growth functions to model and solve real-life problems

Give an example of a real-life situation that is modeled by a logistic growth model.

Example 4: Draw the basic form of the graph of a logistic growth model.

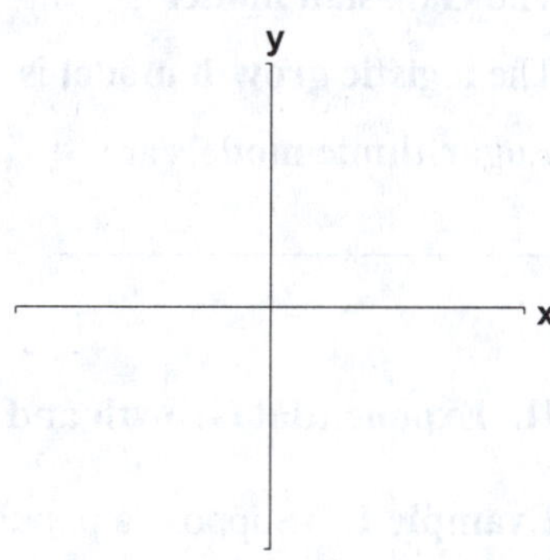

V. Logarithmic Models (Page 264)

What you should learn
How to use logarithmic functions to model and solve real-life problems

Example 5: The number of kitchen widgets y (in millions) demanded each year is given by the model $y = 2 + 3\ \ln(x + 1)$, where $x = 0$ represents the year 2000 and $x \geq 0$. Find the year in which the number of kitchen widgets demanded will be 8.6 million.

VI. Fitting Models to Data (Pages 264–265)

> ***What you should learn***
> How to fit exponential and logarithmic models to sets of data

Describe how to use a graphing utility to fit a logarithmic or an exponential model to data.

An exponential model of the form $y = ab^x$ can be changed to an equivalent exponential model of the form $y = ae^{cx}$ by letting $c =$ ________ in the second model.

Example 6: Rewrite the exponential function $f(x) = 0.2(4)^x$ as a natural logarithmic function.

Example 7: Find an appropriate model, either logarithmic or exponential, for the data in the following table.

x	1	3	5	7	9
y	1.120	2.195	4.303	8.433	16.529

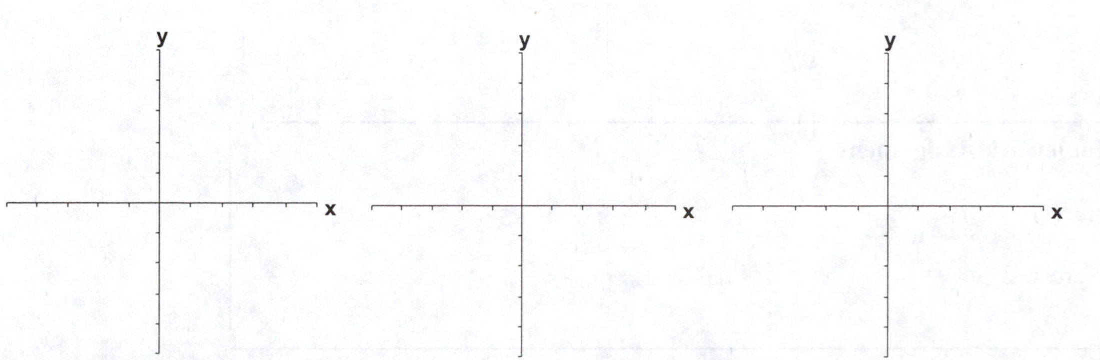

Additional notes

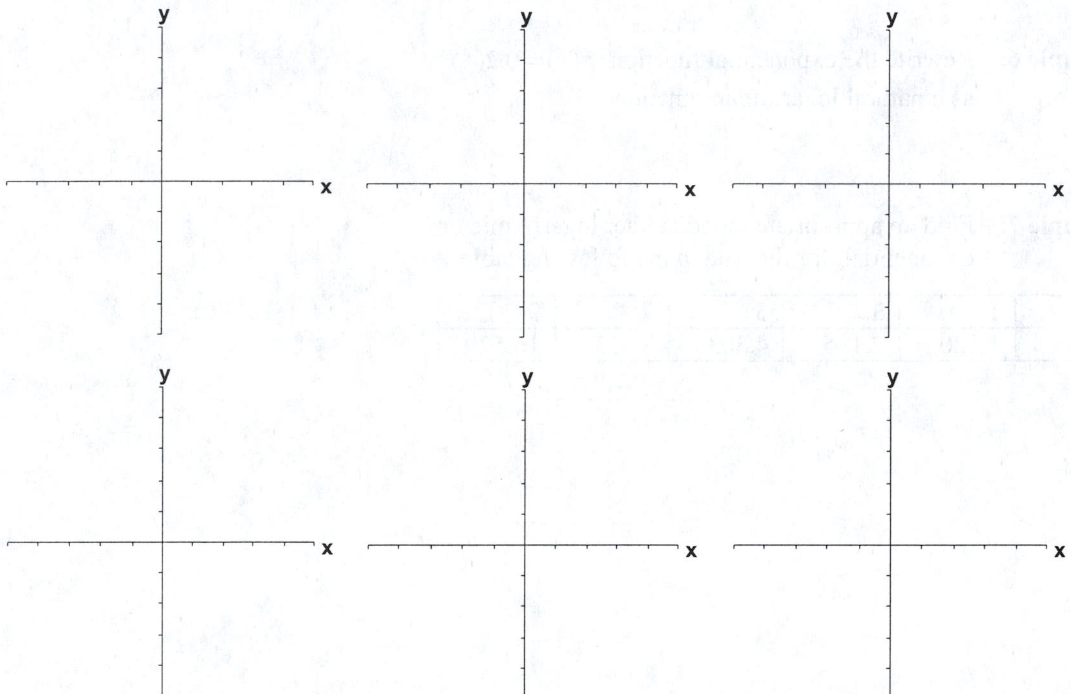

Homework Assignment

Page(s)

Exercises

Chapter 4 Trigonometric Functions

Course Number

Instructor

Date

Section 4.1 Radian and Degree Measure

Objective: In this lesson you learned how to describe an angle and to convert between degree and radian measures.

Important Vocabulary Define each term or concept.

Trigonometry

Central angle (of a circle)

Complementary angles

Supplementary angles

Degree

I. Angles (Page 284)

What you should learn
How to describe angles

An **angle** is determined by . . .

The **initial side** of an angle is . . .

The **terminal side** of an angle is . . .

The **vertex** of an angle is . . .

An angle is in **standard position** when . . .

A **positive angle** is generated by a rotation; whereas a **negative angle** is generated by a rotation.

If two angles are **coterminal,** then they have . . .

II. Radian Measure (Pages 285–287)

What you should learn
How to use radian measure

The measure of an angle is determined by . . .

One **radian** is the measure of a central angle θ that . . .

A central angle of one full revolution (counterclockwise) corresponds to an arc length of $s =$

In general, the radian measure of a central angle θ is obtained by .

A full revolution around a circle of radius r corresponds to an angle of radians. A half revolution around a circle of radius r corresponds to an angle of radians.

Angles with measures between 0 and $\pi/2$ radians are angles. Angles with measures between $\pi/2$ and π radians are angles.

To find an angle that is coterminal to a given angle θ, . . .

Example 1: Find an angle that is coterminal with $\theta = -\pi/8$.

Example 2: Find the supplement of $\theta = \pi/4$.

III. Degree Measure (Pages 287–288)

What you should learn
How to use degree measure

A full revolution (counterclockwise) around a circle corresponds to degrees. A half revolution around a circle corresponds to degrees.

To convert degrees to radians, . . .

To convert radians to degrees, . . .

Example 3: Convert 120° to radians.

Example 4: Convert $9\pi/8$ radians to degrees.

Example 5: Complete the following table of equivalent degree and radian measures for common angles.

θ (degrees)	0°		45°		90°		270°
θ (radians)		$\pi/6$		$\pi/3$		π	

IV. Applications of Angles (Pages 289–290)

What you should learn
How to use angles to model and solve real-life problems

To find the length s of a circular arc of radius r and central angle θ, . . .

Consider a particle moving at constant speed along a circular arc of radius r. If s is the length of the arc traveled in time t, then the **linear speed** of the particle is

linear speed =

If θ is the angle (in radian measure) corresponding to the arc length s, then the **angular speed** of the particle is

angular speed =

Example 6: A 6-inch-diameter gear makes 2.5 revolutions per second. Find the angular speed of the gear in radians per second.

Additional notes

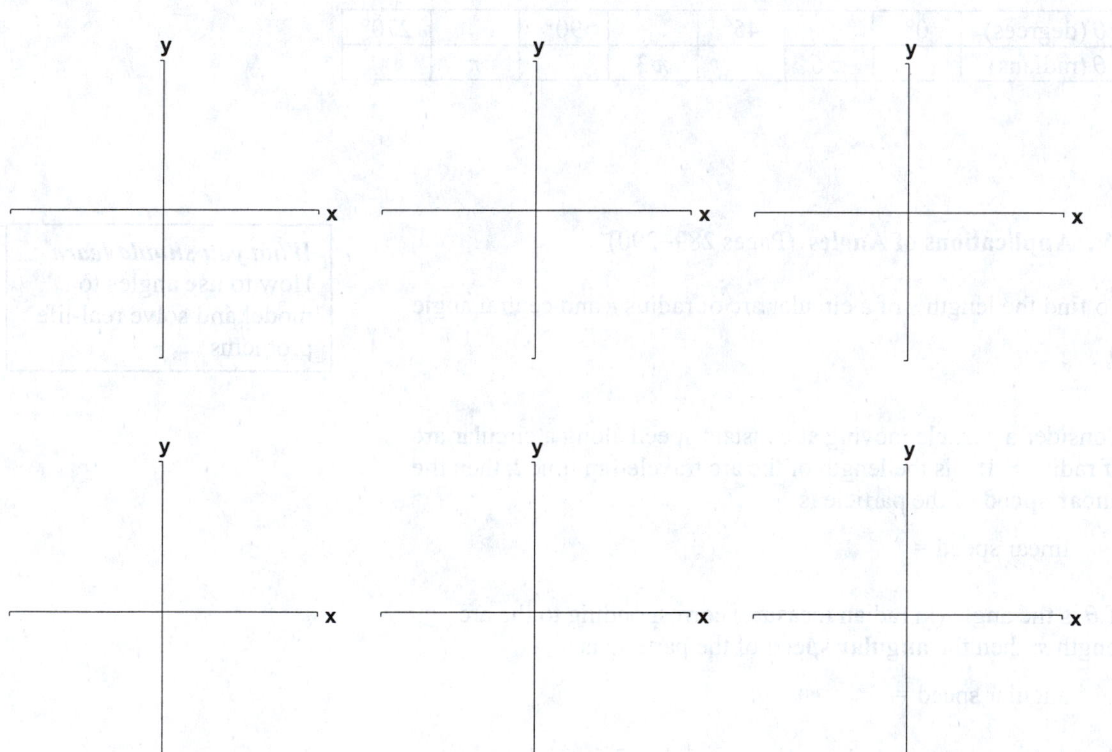

Homework Assignment

Page(s)

Exercises

Course Number

Instructor

Date

Section 4.2 Trigonometric Functions: The Unit Circle

Objective: In this lesson you learned how to identify a unit circle and its relationship to real numbers.

Important Vocabulary Define each term or concept.

Unit circle

Periodic

Period

I. The Unit Circle (Page 295)

What you should learn
How to identify a unit circle and its relationship to real numbers

As the real number line is wrapped around the unit circle, each real number t corresponds to . . .

The real number 2π corresponds to the point on the unit circle.

Each real number t also corresponds to a (in standard position) whose radian measure is t. With this interpretation of t, the arc length formula $s = r\theta$ (with $r = 1$) indicates that . . .

II. The Trigonometric Functions (Pages 296–298)

What you should learn
How to evaluate trigonometric functions using the unit circle

The coordinates x and y are two functions of the real variable t. These coordinates can be used to define six trigonometric functions of t. List the abbreviation for each trigonometric function.

Sine **Cosecant**

Cosine **Secant**

Tangent **Cotangent**

Let t be a real number and let (x, y) be the point on the unit circle corresponding to t. Complete the following definitions of the trigonometric functions:

$\sin t =$ $\cos t =$

$\tan t =$ $\cot t =$

$\sec t =$ $\csc t =$

The cosecant function is the reciprocal of the

function. The cotangent function is the reciprocal of the

function. The secant function is the

reciprocal of the function.

Complete the following table showing the correspondence between the real number t and the point (x, y) on the unit circle when the unit circle is divided into eight equal arcs.

t	0	$\pi/4$	$\pi/2$	$3\pi/4$	π	$5\pi/4$	$3\pi/2$	$7\pi/4$
x								
y								

Complete the following table showing the correspondence between the real number t and the point (x, y) on the unit circle when the unit circle is divided into 12 equal arcs.

t	0	$\pi/6$	$\pi/3$	$\pi/2$	$2\pi/3$	$5\pi/6$	π	$7\pi/6$	$4\pi/3$	$3\pi/2$	$5\pi/3$	$11\pi/6$
x												
y												

Example 1: Find the following:

(a) $\cos\frac{\pi}{3}$ (b) $\tan\frac{3\pi}{4}$ (c) $\csc\frac{7\pi}{6}$

III. Domain and Period of Sine and Cosine (Pages 298–299)

What you should learn
How to use the domain and period to evaluate sine and cosine functions

The sine function's domain is

and its range is

The cosine function's domain is ,

and its range is .

The period of the sine function is . The

period of the cosine function is .

Which trigonometric functions are even functions?

Which trigonometric functions are odd functions?

Example 2: Evaluate $\sin \frac{31\pi}{6}$

IV. Evaluating Trigonometric Functions with a Calculator (Page 299)

What you should learn
How to use a calculator to evaluate trigonometric functions

To evaluate the secant function with a calculator, . . .

Example 3: Use a calculator to evaluate (a) $\tan 4\pi/3$, and (b) $\cos 3$.

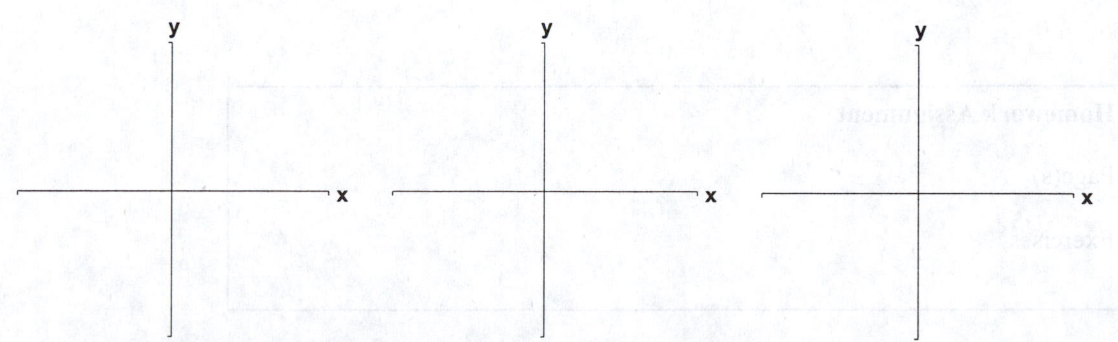

Additional notes

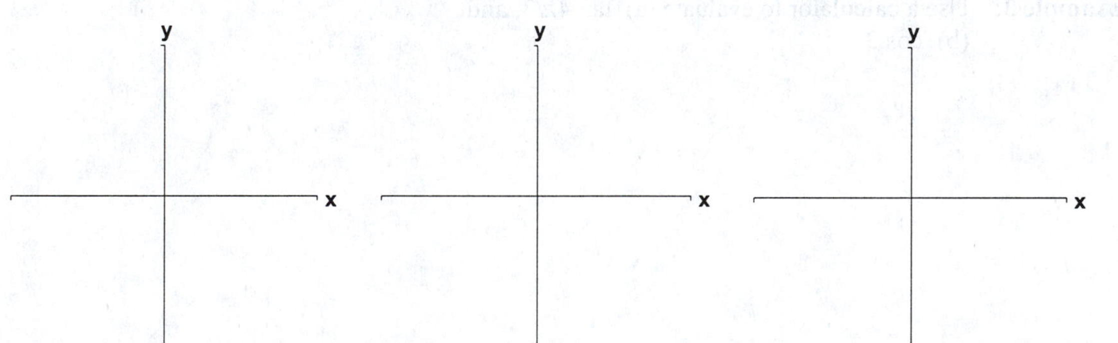

Homework Assignment

Page(s)

Exercises

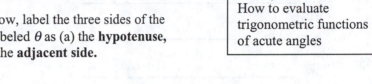

Course Number

Instructor

Date

Section 4.3 Right Triangle Trigonometry

Objective: In this lesson you learned how to evaluate trigonometric functions of acute angles and how to use the fundamental trigonometric identities.

I. The Six Trigonometric Functions (Pages 303–305)

What you should learn
How to evaluate trigonometric functions of acute angles

In the right triangle shown below, label the three sides of the triangle relative to the angle labeled θ as (a) the **hypotenuse,** (b) the **opposite side,** and (c) the **adjacent side.**

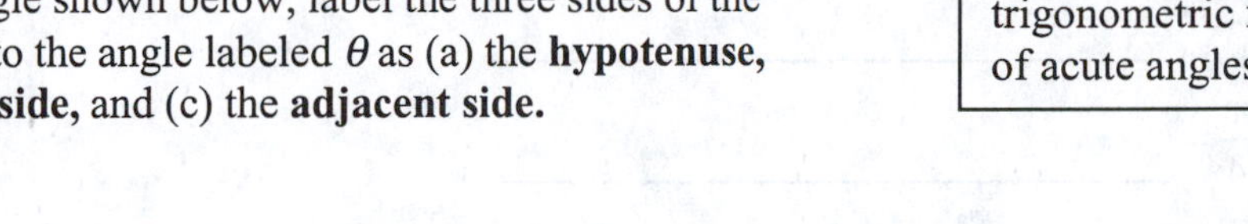

θ

Let θ be an acute angle of a right triangle. Define the six trigonometric functions of the angle θ using opp = the length of the side opposite θ, adj = the length of the side adjacent to θ, and hyp = the length of the hypotenuse.

$\sin \theta =$ $\cos \theta =$

$\tan \theta =$ $\csc \theta =$

$\sec \theta =$ $\cot \theta =$

The cosecant function is the reciprocal of the function. The cotangent function is the reciprocal of the function. The secant function is the reciprocal of the function.

Example 1: In the right triangle below, find $\sin \theta$, $\cos \theta$, and $\tan \theta$.

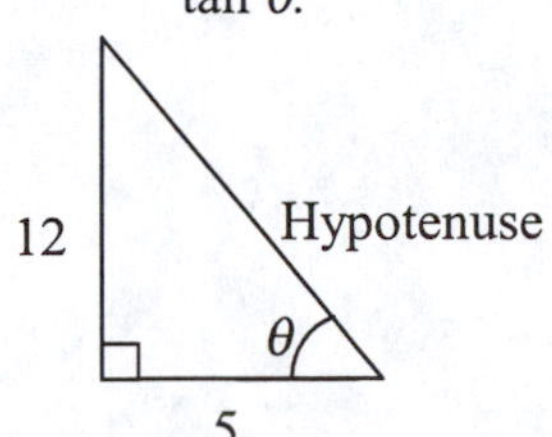

Give the sines, cosines, and tangents of the following special angles:

$\sin 30° = \sin\dfrac{\pi}{6} =$ ______________________

$\cos 30° = \cos\dfrac{\pi}{6} =$ ______________________

$\tan 30° = \tan\dfrac{\pi}{6} =$ ______________________

$\sin 45° = \sin\dfrac{\pi}{4} =$ ______________________

$\cos 45° = \cos\dfrac{\pi}{4} =$ ______________________

$\tan 45° = \tan\dfrac{\pi}{4} =$ ______________________

$\sin 60° = \sin\dfrac{\pi}{3} =$ ______________________

$\cos 60° = \cos\dfrac{\pi}{3} =$ ______________________

$\tan 60° = \tan\dfrac{\pi}{3} =$ ______________________

Cofunctions of complementary angles are ______________. If θ is an acute angle, then:

$\sin(90° - \theta) =$ $\cos(90° - \theta) =$

$\tan(90° - \theta) =$ $\cot(90° - \theta) =$

$\sec(90° - \theta) =$ $\csc(90° - \theta) =$

II. Trigonometric Identities (Pages 306–307)

> ***What you should learn***
> How to use the fundamental trigonometric identities

List six reciprocal identities:

1)

2)

3)

4)

5)

6)

List two quotient identities:

1)

2)

List three Pythagorean identities:

1)

2)

3)

III. Evaluating Trigonometric Functions with a Calculator (Page 307)

What you should learn
How to use a calculator to evaluate trigonometric functions

To use a calculator to evaluate trigonometric functions of angles measured in degrees, . . .

Example 2: Use a calculator to evaluate (a) tan 35.4°, and (b) cos 3.25°

IV. Applications Involving Right Triangles (Pages 307–309)

What you should learn
How to use trigonometric functions to model and solve real-life problems

What does it mean to "solve a right triangle?"

An **angle of elevation** is . . .

An **angle of depression** is . . .

Describe a real-life situation in which solving a right triangle would be appropriate or useful.

Additional notes

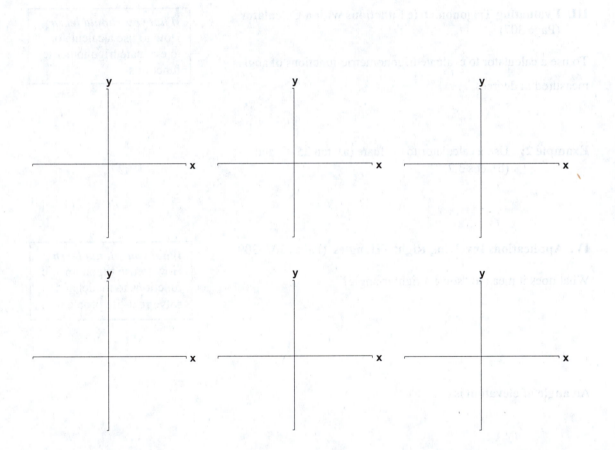

y x

y x

y x

y x

y x

y x

Homework Assignment

Page(s)

Exercises

Course Number
Instructor
Date

Section 4.4 Trigonometric Functions of Any Angle

Objective: In this lesson you learned how to evaluate trigonometric functions of any angle.

Important Vocabulary	Define each term or concept.
Reference angles	

I. Introduction (Pages 314–315)

What you should learn
How to evaluate trigonometric functions of any angle

Let θ be an angle in standard position with (x, y) a point on the terminal side of θ and $r = \sqrt{x^2 + y^2} \neq 0$. Complete the following definitions of the trigonometric functions of any angle:

$\sin\theta =$ $\cos\theta =$

$\tan\theta =$ $\cot\theta =$

$\sec\theta =$ $\csc\theta =$

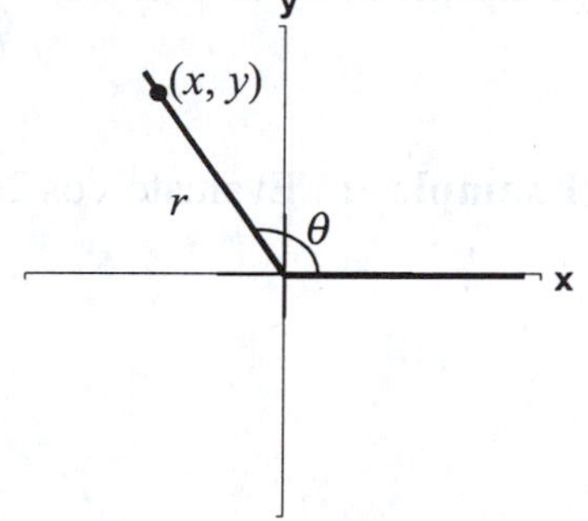

Name the quadrants in which the sine function is positive.

Name the quadrants in which the sine function is negative.

Name the quadrants in which the cosine function is positive.

Name the quadrants in which the cosine function is negative.

Name the quadrants in which the tangent function is positive.

Name the quadrants in which the tangent function is negative.

Example 1: If $\sin\theta = \frac{1}{2}$ and $\tan\theta < 0$, find $\cos\theta$.

II. Reference Angles (Page 316)

> ***What you should learn***
> How to use reference angles to evaluate trigonometric functions

Example 2: Find the reference angle θ' for
(a) $\theta = 210°$ (b) $\theta = 4.1$

III. Trigonometric Functions of Real Numbers
(Pages 317–319)

> ***What you should learn***
> How to evaluate trigonometric functions of real numbers

To find the value of a trigonometric function of any angle θ, . . .

Example 3: Evaluate $\sin \dfrac{11\pi}{6}$.

Example 4: Evaluate $\cos 240°$.

y x

y x

y x

Homework Assignment

Page(s)

Exercises

Course Number

Instructor

Date

Section 4.5 Graphs of Sine and Cosine Functions

Objective: In this lesson you learned how to sketch the graphs of sine and cosine functions and translations of these functions.

Important Vocabulary Define each term or concept.

Sine curve

Amplitude

Phase shift

I. Basic Sine and Cosine Curves (Pages 323–324)

What you should learn
How to sketch the graphs of basic sine and cosine functions

For $0 \leq x \leq 2\pi$, the sine function has its maximum point at

, its minimum point at ,

and its intercepts at

For $0 \leq x \leq 2\pi$, the cosine function has its maximum points at

, its minimum point at ,

and its intercepts at .

Example 1: Sketch the basic sine curve on the interval $[0, 2\pi]$.

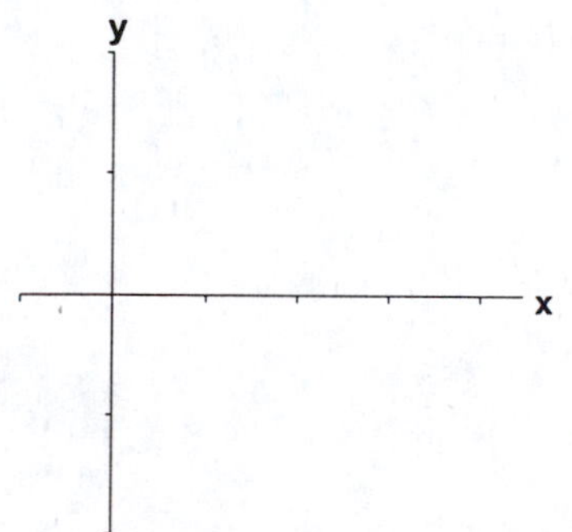

Example 2: Sketch the basic cosine curve on the interval $[0, 2\pi]$.

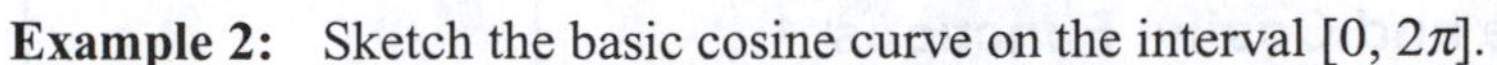

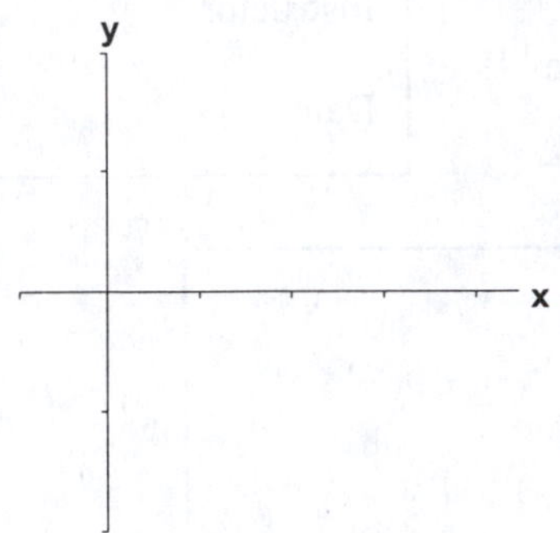

II. Amplitude and Period of Sine and Cosine Curves
(Pages 325–326)

> ***What you should learn***
> How to use amplitude and period to help sketch the graphs of sine and cosine functions

The constant factor a in $y = a \sin x$ acts as . . .

If $|a| > 1$, the basic sine curve is ________________. If $|a| < 1$, the basic sine curve is ________________. The result is that the graph of $y = a \sin x$ ranges between ________________ instead of between -1 and 1. The absolute value of a is the ________________ of the function $y = a \sin x$.

The graph of $y = 0.5 \sin x$ is a(n) ________________ in the x-axis of the graph of $y = -0.5 \sin x$.

Let b be a positive real number. The **period** of $y = a \sin bx$ and $y = a \cos bx$ is ________________. If $0 < b < 1$, the period of $y = a \sin bx$ is ________________ than 2π and represents a ________________ of the graph of $y = a \sin bx$.

If $b > 1$, the period of $y = a \sin bx$ is ________________ than 2π and represents a ________________ of the graph of $y = a \sin bx$.

Example 3: Find the amplitude and the period of $y = -4\cos 3x$.

Example 4: Find the five key points (intercepts, maximum points, and minimum points) of the graph of $y = -4\cos 3x$.

III. Translations of Sine and Cosine Curves (Pages 327–328)

What you should learn
How to sketch translations of graphs of sine and cosine functions

The constant c in the general equations $y = a\sin(bx - c)$ and $y = a\cos(bx - c)$ creates . . .

Comparing $y = a\sin bx$ with $y = a\sin(bx - c)$, the graph of $y = a\sin(bx - c)$ completes one cycle from to . By solving for x, the interval for one cycle is found to be to . This implies that the graph of $y = a\sin(bx - c)$ is the graph of $y = a\sin bx$ shifted by the amount .

The period of the graph of $y = a\cos(bx - c)$ is .

Example 5: Find the amplitude, period, and phase shift of $y = 2\sin(x - \pi/4)$.

Example 6: Find the five key points (intercepts, maximum points, and minimum points) of the graph of $y = 2\sin(x - \pi/4)$.

The constant d in the equation $y = d + a\sin(bx - c)$ causes a(n) . For $d > 0$, the shift is . For $d < 0$, the shift is .

The graph oscillates about .

IV. Mathematical Modeling (Page 329)

What you should learn
How to use sine and cosine functions to model real-life data

Describe a real-life situation which can be modeled by a sine or cosine function.

Example 7: Find a trigonometric function to model the data in the following table.

x	0	$\pi/2$	π	$3\pi/2$	2π
y	2	4	2	0	2

Additional notes

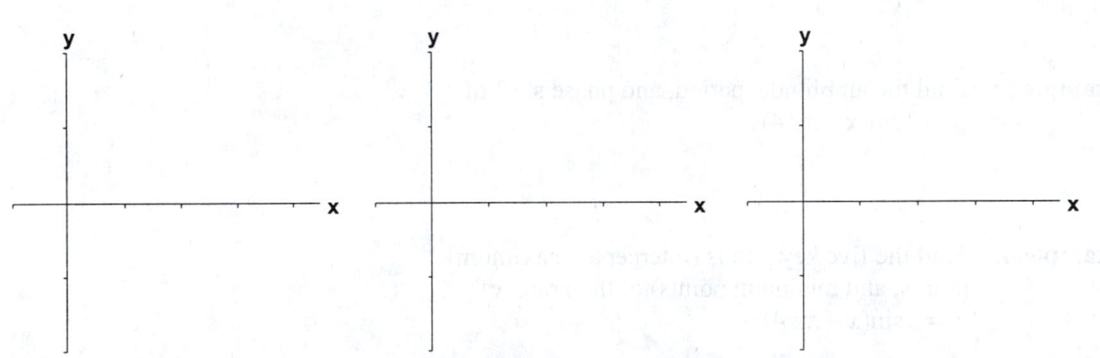

Homework Assignment

Page(s)

Exercises

Course Number

Instructor

Date

Section 4.6 Graphs of Other Trigonometric Functions

Objective: In this lesson you learned how to sketch the graphs of other trigonometric functions.

Important Vocabulary	Define each term or concept.
Damping factor	

I. Graph of the Tangent Function (Pages 334–335)

What you should learn
How to sketch the graphs of tangent functions

Because the tangent function is odd, the graph of $y = \tan x$ is symmetric with respect to the ________ . The period of the tangent function is ________ . On the interval [0, π], the tangent function is undefined, and thus has a vertical asymptote, at $x =$ ________ . The domain of the tangent function is ________ , and the range of the tangent function is ________ .

Describe how to sketch the graph of a function of the form $y = a\tan(bx - c)$.

II. Graph of the Cotangent Function (Page 336)

What you should learn
How to sketch the graphs of cotangent functions

The graph of $y = \cot x$ is symmetric with respect to the ________ . The period of the cotangent function is ________ . On the interval (0, π], the cotangent function is undefined, and thus has a vertical asymptote, at $x =$ ________ .

Larson/Hostetler/Edwards *Precalculus Functions and Graphs: A Graphing Approach, Third Edition*
Larson/Hostetler/Edwards *Precalculus with Limits: A Graphing Approach, Third Edition*
Student Success Organizer

The domain of the cotangent function is ,

and the range of the cotangent function is .

III. Graphs of the Reciprocal Functions (Pages 337–338)

What you should learn
How to sketch the graphs of secant and cosecant functions

At a given value of x, the y-coordinate of csc x is the reciprocal of the y-coordinate of .

The graph of $y = \csc x$ is symmetric with respect to the . The period of the cosecant function is . On the interval $(0, \pi]$, the cosecant function is undefined, and thus has a vertical asymptote, at $x =$.

The domain of the cosecant function is ,

and the range of the cosecant function is .

At a given value of x, the y-coordinate of sec x is the reciprocal of the y-coordinate of .

The graph of $y = \sec x$ is symmetric with respect to the . The period of the secant function is .

On the interval $[0, \pi]$, the secant function is undefined, and thus has a vertical asymptote, at $x =$. The domain of the secant function is , and the range of the secant function is .

To sketch the graph of a secant or cosecant function, . . .

What you should learn
How to sketch the graphs of damped trigonometric functions

IV. Damped Trigonometric Graphs (Pages 339–340)

Explain how to sketch the graph of the damped trigonometric function $y = f(x)\cos x$, where $f(x)$ is the damping factor.

Additional notes

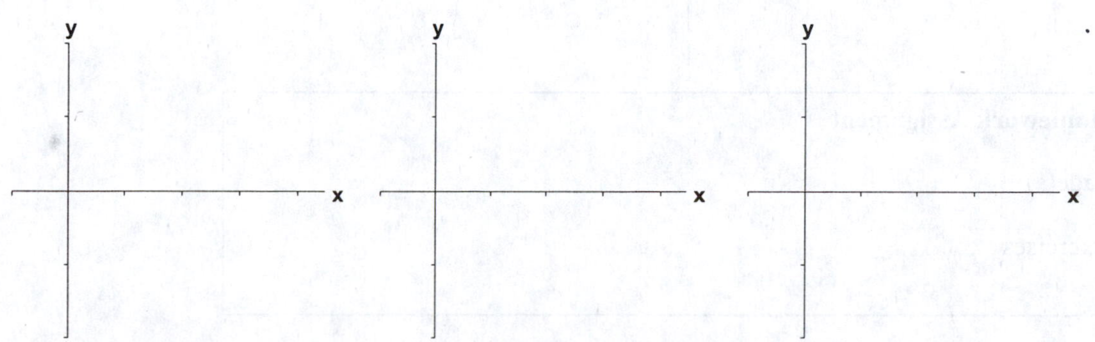

Additional notes

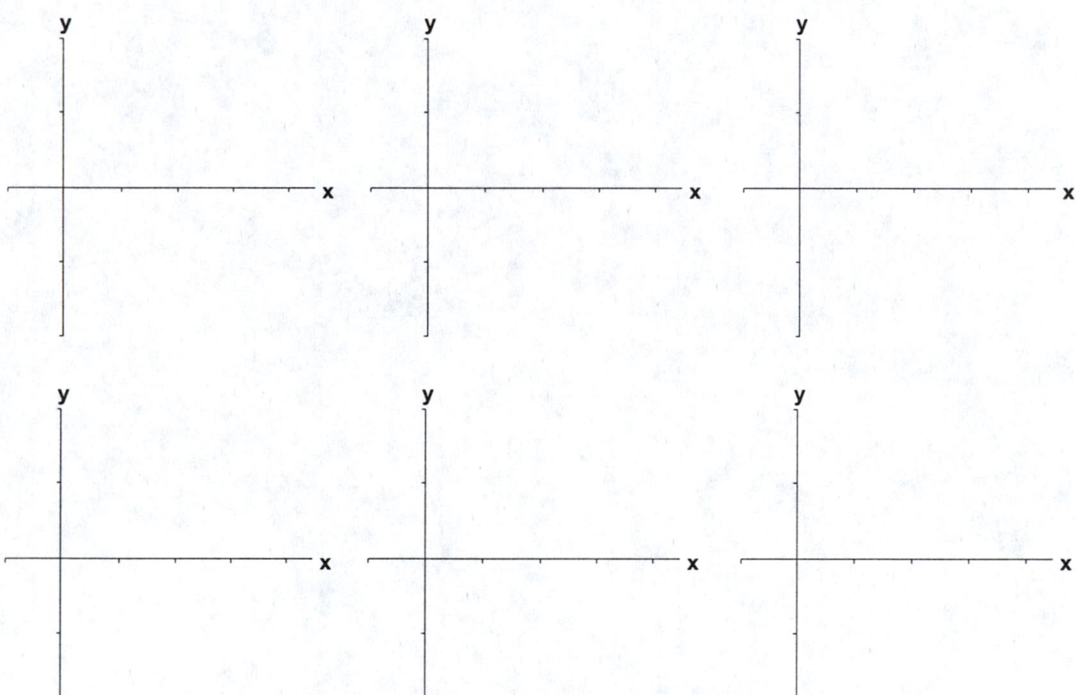

Homework Assignment

Page(s)

Exercises

Course Number
Instructor
Date

Section 4.7 Inverse Trigonometric Functions

Objective: In this lesson you learned how to evaluate the inverse trigonometric functions and how to evaluate compositions of trigonometric functions.

I. Inverse Sine Function (Pages 345–346)

> ***What you should learn***
> How to evaluate the inverse sine function

The **inverse sine function** is defined by . . .

The domain of $y = \arcsin x$ is . The range of $y = \arcsin x$ is .

Example 1: Find the exact value: $\arcsin(-1)$.

II. Other Inverse Trigonometric Functions (Pages 347–348)

> ***What you should learn***
> How to evaluate the other inverse trigonometric functions

The **inverse cosine function** is defined by . . .

The domain of $y = \arccos x$ is . The range of $y = \arccos x$ is .

Example 2: Find the exact value: $\arccos \frac{1}{2}$.

The **inverse tangent function** is defined by . . .

The domain of $y = \arctan x$ is . The range of $y = \arctan x$ is .

Example 3: Find the exact value: $\arctan(\sqrt{3})$.

Example 4: Use a calculator to approximate the value (if possible). Round to four decimal places.
(a) arcos 0.85 (b) arcsin 3.1415

III. Compositions of Functions (Pages 349–350)

What you should learn
How to evaluate the compositions of trigonometric functions

State the Inverse Property for the Sine function.

State the Inverse Property for the Cosine function.

State the Inverse Property for the Tangent function.

Example 5: If possible, find the exact value:
(a) arcsin(sin $3\pi/4$) (b) cos(arccos 0)

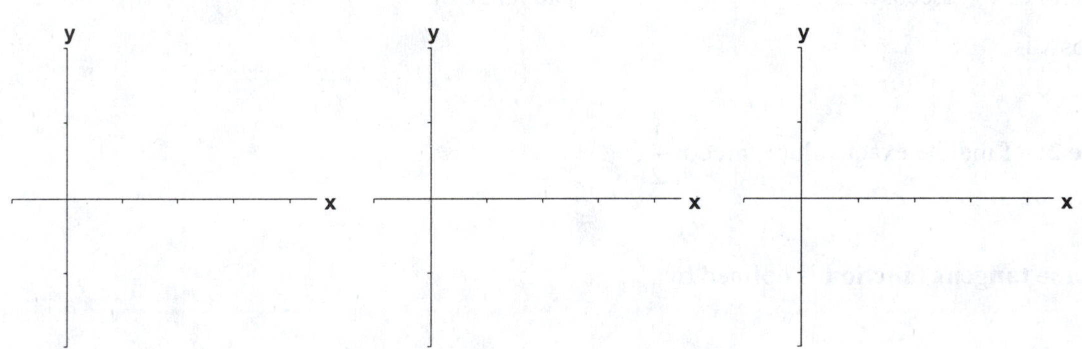

Homework Assignment

Page(s)

Exercises

Course Number

Instructor

Date

Section 4.8 Applications and Models

Objective: In this lesson you learned how to use trigonometric functions to solve real-life problems.

I. Applications Involving Right Triangles (Pages 355–356)

What you should learn
How to solve real-life problems involving right triangles

Example 1: A ladder leaning against a house reaches 24 feet up the side of the house. The ladder makes a 60° angle with the ground. How far is the base of the ladder from the house? Round your answer to two decimal places.

II. Trigonometry and Bearings (Page 357)

What you should learn
How to solve real-life problems involving directional bearings

Used to give directions in surveying and navigation, a **bearing** measures . . .

The bearing N 70° E means . . .

Example 2: Write the bearing for the path shown in the diagram at the right.

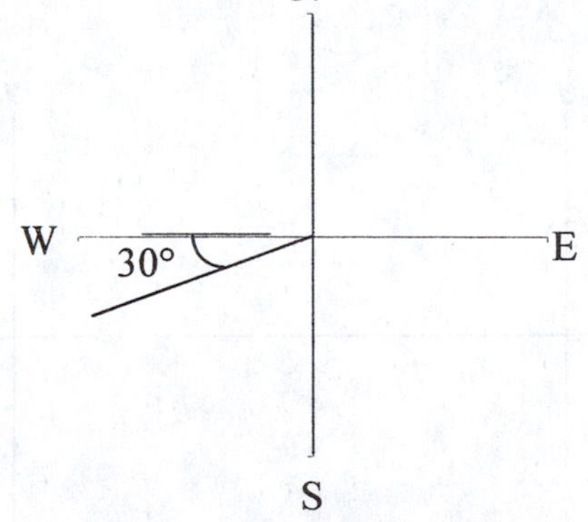

III. Harmonic Motion (Pages 358–360)

What you should learn
How to solve real-life problems involving harmonic motion

The vibration, oscillation, or rotation of an object under ideal conditions such that the object's uniform and regular motion can be described by a sine or cosine function is called

.

Larson/Hostetler/Edwards *Precalculus Functions and Graphs: A Graphing Approach, Third Edition*
Larson/Hostetler/Edwards *Precalculus with Limits: A Graphing Approach, Third Edition*
Student Success Organizer

A point that moves on a coordinate line is said to be in simple harmonic motion if . . .

The simple harmonic motion has amplitude , period , and frequency .

Example 3: Given the equation for simple harmonic motion $d = 3\sin\frac{t}{2}$, find:

(a) the maximum displacement,
(b) the frequency of the simple harmonic motion, and
(c) the period of the simple harmonic motion.

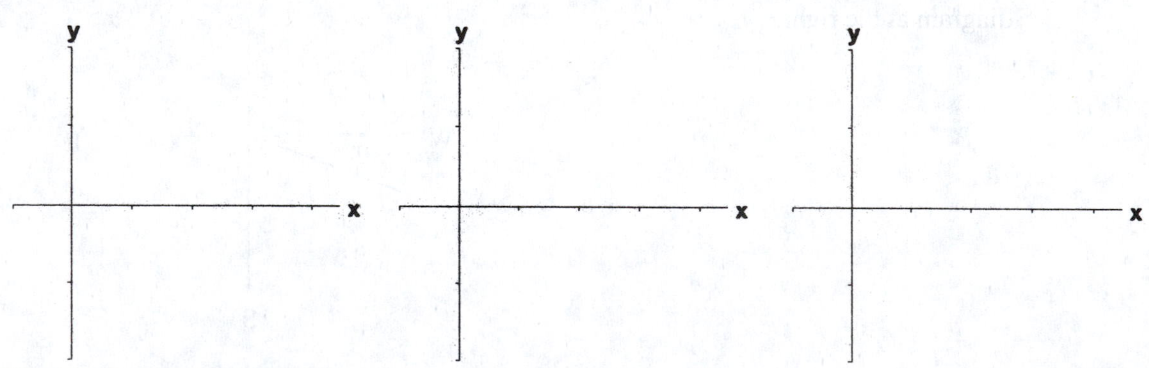

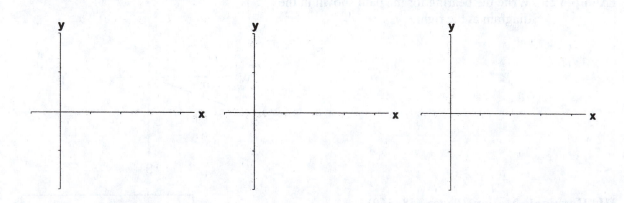

Homework Assignment

Page(s)

Exercises

Chapter 5 Analytic Trigonometry

Course Number

Instructor

Date

Section 5.1 Using Fundamental Identities

Objective: In this lesson you learned how to use fundamental trigonometric identities to evaluate trigonometric functions and simplify trigonometric expressions.

What you should learn
How to recognize and write the fundamental trigonometric identities

I. Introduction (Page 376)

Name four ways in which the fundamental trigonometric identities can be used:

1)

2)

3)

4)

The Fundamental Trigonometric Identities

List six reciprocal identities:

1)

2)

3)

4)

5)

6)

List six cofunction identities:

1)

2)

3)

4)

5)

6)

List two quotient identities:

1)

2)

List three Pythagorean identities:

1)

2)

3)

List six even/odd identities:

1)

2)

3)

4)

5)

6)

II. Using the Fundamental Identities (Pages 377–380)

What you should learn
How to use the fundamental trigonometric identities to evaluate trigonometric functions, simplify trigonometric expressions, and rewrite trigonometric expressions

Example 1: Explain how to use the fundamental trigonometric identities to find the value of tan u given that $\sec u = 2$.

Example 2: Explain how to use the fundamental trigonometric identities to simplify $\sec x - \tan x \sin x$.

Example 3: Explain how to use a graphing utility to verify whether $\sec x \sin^3 x + \sin x \cos x = \tan x$ is an identity.

Homework Assignment

Page(s)

Exercises

Course Number

Instructor

Date

Section 5.2 Verifying Trigonometric Identities

Objective: In this lesson you learned how to verify trigonometric identities.

I. Introduction (Page 384)

What you should learn
How to understand the difference between conditional equations and identities

The key to verifying identities is . . .

An identity is . . .

II. Verifying Trigonometric Identities (Pages 384–388)

What you should learn
How to verify trigonometric identities

Complete the following list of guidelines for verifying trigonometric identities:

1)

2)

3)

4)

5)

Example 1: Describe a strategy for verifying the identity $\sin\theta\tan\theta + \cos\theta = \sec\theta$. Then verify the identity.

Example 2: Describe a strategy for verifying the identity $\sin^2 x(\csc x - 1)(\csc x + 1) = 1 - \sin^2 x$. Then verify the identity.

Example 3: Verify the identity $\cot^5 \alpha = \cot^3 \alpha \csc^2 \alpha - \cot^3 \alpha$.

Additional notes

Homework Assignment

Page(s)

Exercises

Course Number

Instructor

Date

Section 5.3 Solving Trigonometric Equations

Objective: In this lesson you learned how to use standard algebraic techniques and inverse trigonometric functions to solve trigonometric equations.

I. Introduction (Pages 392–394)

What you should learn
How to use standard algebraic techniques to solve trigonometric equations

To solve a trigonometric equation, . . .

The preliminary goal in solving trigonometric equations is . . .

How many solutions does the equation $\sec x = 2$ have? Explain.

Example 1: Solve $2\cos^2 x - 1 = 0$.

To solve an equation in which two or more trigonometric functions occur, . . .

II. Equations of Quadratic Type (Pages 394–396)

What you should learn
How to solve trigonometric equations of quadratic type

Give an example of a trigonometric equation of quadratic type.

To solve a trigonometric equation of quadratic type, . . .

Example 2: Solve $\tan^2 x + 2\tan x = -1$.

Care must be taken when squaring both sides of a trigonometric equation to obtain a quadratic because . . .

III. Functions Involving Multiple Angles (Page 397)

What you should learn
How to solve trigonometric equations involving multiple angles

Give an example of a trigonometric function of multiple angles.

Example 3: Solve $\sin 4x = \frac{\sqrt{2}}{2}$.

IV. Using Inverse Functions (Page 398–399)

What you should learn
How to use inverse trigonometric functions to solve trigonometric equations

Example 4: Use inverse functions to solve the equation $\tan^2 x + 4\tan x + 4 = 0$.

Homework Assignment

Page(s)

Exercises

Course Number

Instructor

Date

Section 5.4 Sum and Difference Formulas

Objective: In this lesson you learned how to use sum and difference formulas to rewrite and evaluate trigonometric functions.

What you should learn
How to use sum and difference formulas to evaluate trigonometric functions, to verify identities, and to solve trigonometric equations

I. Using Sum and Difference Formulas (Pages 404–407)

List the sum and difference formulas for sine, cosine, and tangent.

Example 1: Use a sum or difference formula to find the exact value of tan 255°.

Example 2: Find the exact value of cos 95° cos 35° + sin 95° sin 35°.

A **reduction formula** is . . .

Example 3: Derive a reduction formula for $\sin\left(t+\frac{\pi}{2}\right)$.

Example 4: Find all solutions of $\cos(x - \frac{\pi}{3}) + \cos(x + \frac{\pi}{3}) = 1$ in the interval $[0, 2\pi)$.

Additional notes

Homework Assignment

Page(s)

Exercises

Course Number

Instructor

Date

Section 5.5 Multiple-Angle and Product-Sum Formulas

Objective: In this lesson you learned how to use multiple-angle formulas, power-reducing formulas, half-angle formulas, and product-sum formulas to rewrite and evaluate trigonometric functions.

I. Multiple-Angle Formulas (Pages 411–413)

What you should learn
How to use multiple-angle formulas to rewrite and evaluate trigonometric functions

The most commonly used multiple-angle formulas are the

, which are listed below:

$\sin 2u =$

$\cos 2u =$

$=$

$=$

$\tan 2u =$

To obtain other multiple-angle formulas, . . .

Example 1: Use multiple-angle formulas to express $\cos 3x$ in terms of $\cos x$.

II. Power-Reducing Formulas (Page 413)

What you should learn
How to use power-reducing formulas to rewrite and evaluate trigonometric functions

Power-reducing formulas can be used to . . .

The power-reducing formulas are:

$\sin^2 u =$

$\cos^2 u =$

$\tan^2 u =$

III. Half-Angle Formulas (Pages 414–415)

> ***What you should learn***
> How to use half-angle formulas to rewrite and evaluate trigonometric functions

List the **half-angle formulas:**

$\sin \frac{u}{2} =$

$\cos \frac{u}{2} =$

$\tan \frac{u}{2} = \qquad\qquad =$

The signs of sin (u/2) and cos (u/2) depend on . . .

Example 2: Find the exact value of tan 15°.

IV. Product-to-Sum Formulas (Pages 415–417)

> ***What you should learn***
> How to use product-sum formulas to rewrite and evaluate trigonometric functions

The **product-to-sum formulas** can be used to . . .

The product-to-sum formulas are:

$\sin u \sin v =$

$\cos u \cos v =$

$\sin u \cos v =$

$\cos u \sin v =$

Larson/Hostetler/Edwards *Precalculus Functions and Graphs: A Graphing Approach, Third Edition*
Larson/Hostetler/Edwards *Precalculus with Limits: A Graphing Approach, Third Edition*
Student Success Organizer

Example 3: Write $\cos 3x \cos 2x$ as a sum or difference.

The **sum-to-product formulas** can be used to . . .

The sum-to-product formulas are:

$\sin x + \sin y =$

$\sin x - \sin y =$

$\cos x + \cos y =$

$\cos x - \cos y =$

Example 4: Write $\cos 4x + \cos 2x$ as a sum or difference.

Additional notes

Additional notes

Homework Assignment

Page(s)

Exercises

Chapter 6 Additional Topics in Trigonometry

Course Number
Instructor
Date

Section 6.1 Law of Sines

Objective: In this lesson you learned how to use the Law of Sines to solve oblique triangles and how to find the areas of oblique triangles.

Important Vocabulary	Define each term or concept.
Oblique triangle	

I. Introduction (Pages 428–429)

What you should learn
How to use the Law of Sines to solve oblique triangles (AAS or ASA)

State the **Law of Sines.**

To solve an oblique triangle, you need to know the measure of at least one side and any two other parts of the triangle. Describe two cases that can be solved using the Law of Sines.

Example 1: For the triangle shown at the right, $A = 31.6°$, $C = 42.9°$, and $a = 10.4$ meters. Find the length of side c.

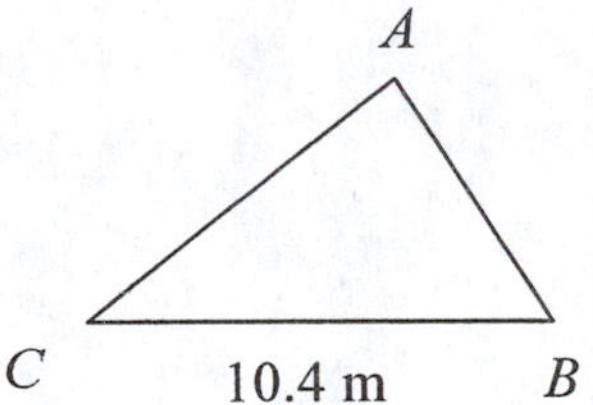

II. The Ambiguous Case (SSA) (Pages 430–431)

What you should learn
How to use the Law of Sines to solve oblique triangles (SSA)

If two sides and one opposite angle of an oblique triangle are given, possible situations can occur, which are:

Example 2: For a triangle having $A = 25°$, $b = 54$ feet, and $a = 26$ feet, how many solutions are possible?

Example 3: For the triangle shown at the right, $A = 110°$, $c = 16$ centimeters, and $a = 25$ centimeters. Find the length of side b.

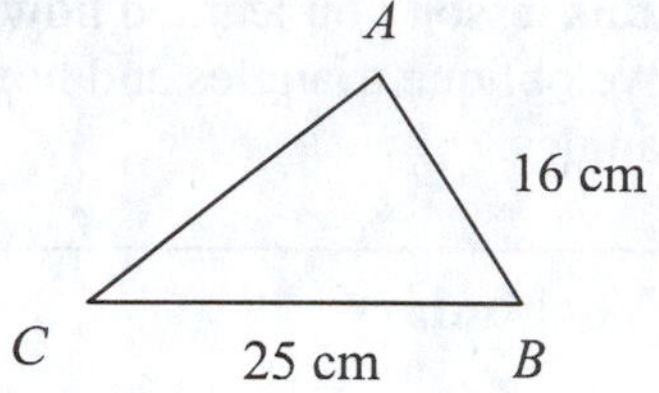

III. Area of an Oblique Triangle (Page 432)

What you should learn
How to find the areas of oblique triangles

The area of any triangle is ______ the product of the lengths of two sides times the sine of ______.

That is,

Area =

Example 4: Find the area of a triangle having two sides of lengths 30 feet and 48 feet and an included angle of 40°.

IV. Applications of the Law of Sines (Page 433)

What you should learn
How to use the Law of Sines to model and solve real-life problems

Describe a real-life situation in which the Law of Sines could be used.

Homework Assignment

Page(s)

Exercises

Course Number

Instructor

Date

Section 6.2 Law of Cosines

Objective: In this lesson you learned how to use the Law of Cosines to solve oblique triangles and to use Heron's Formula to find the area of a triangle.

I. Introduction (Pages 437–439)

What you should learn
How to use the Law of Cosines to solve oblique triangles (SSS or SAS) and to use the Law of Cosines to model and solve real-life problems

State the **Law of Cosines**.

Example 1: Using the triangle shown at the right, find angle *A*.

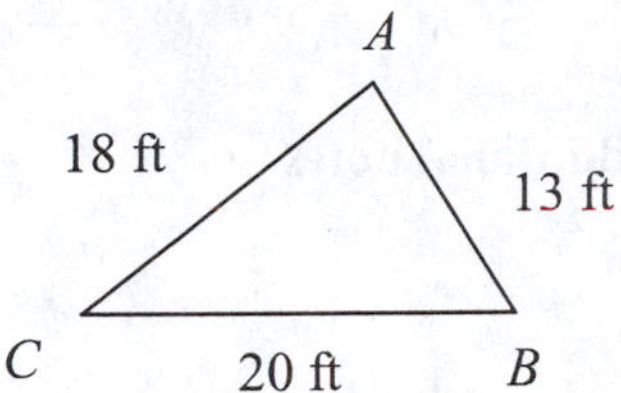

When given the lengths of all three sides of a triangle and asked to find all three angles, which angle should be found first? Why?

Example 2: In the triangle shown at the right, if $A = 62°$, find the length of side *a*.

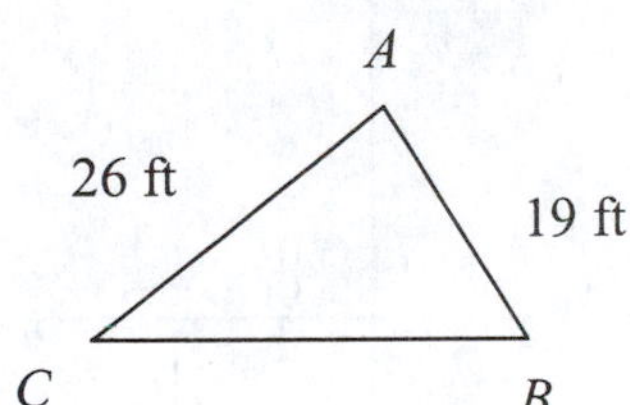

Describe a real-life situation in which the Law of Cosines could be used.

II. Heron's Area Formula (Page 440)

What you should learn
How to use Heron's Area Formula to find the area of a triangle

Heron's Area Formula states that given any triangle with sides of length a, b, and c, the area of the triangle is:

Area = $\sqrt{\quad}$ ______________________

where s = ______________________ .

Example 3: Find the area of a triangle having sides of length $a = 14$ cm, $b = 21$ cm, and $c = 27$ cm.

Additional notes

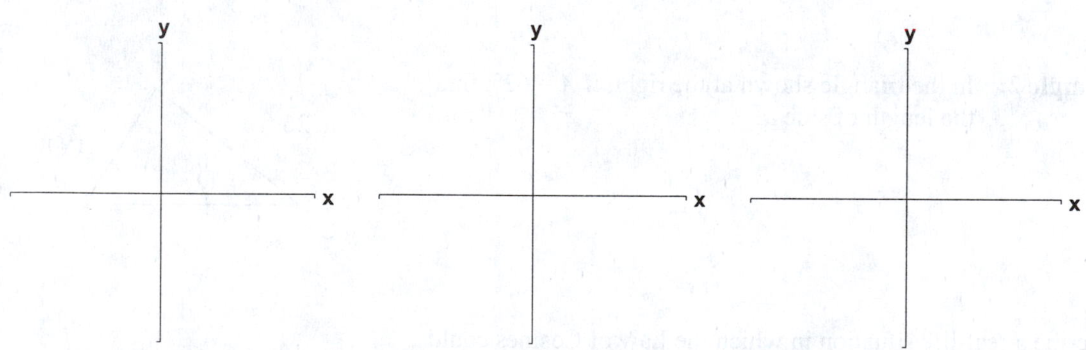

Homework Assignment

Page(s)

Exercises

Course Number

Instructor

Date

Section 6.3 Vectors in the Plane

Objective: In this lesson you learned how to write the component forms of vectors, perform basic vector operations, and find the direction angles of vectors.

Important Vocabulary Define each term or concept.

Vector v in the plane

Standard position

Zero vector

Unit vector

Standard unit vectors

Direction angle

I. Introduction (Page 444)

What you should learn
How to represent vectors as directed line segments

A **directed line segment** has an ____________ and a ____________.

The magnitude of the directed line segment $\overrightarrow{PQ}$, denoted by ____________, is its ____________. The magnitude of a directed line segment can be found by . . .

II. Component Form of a Vector (Page 445)

What you should learn
How to write the component forms of vectors

A vector whose initial point is at the origin (0, 0) can be uniquely represented by the coordinates of its terminal point (v_1, v_2). This is the ____________, written $\mathbf{v} = \langle v_1, v_2 \rangle$, where v_1 and v_2 are the ____________ of **v**.

The component form of the vector with initial point $P = (p_1, p_2)$ and terminal point $Q = (q_1, q_2)$ is
$\overrightarrow{PQ} =$ ____________ $=$ ____________ $= \mathbf{v}$.

The **magnitude** (or length) **of v** is:

$\|\mathbf{v}\| = \sqrt{\qquad\qquad} = \sqrt{\qquad}$

Example 1: Find the component form and magnitude of the vector **v** that has (1, 7) as its initial point and (4, 3) as its terminal point.

III. Vector Operations (Pages 446–448)

What you should learn
How to perform basic vector operations and represent them graphically

Geometrically, the product of a vector **v** and a scalar *k* is . . .

If *k* is positive, *k***v** has the ________ direction as **v**, and if *k* is negative, *k***v** has the ________ direction.

To add two vectors geometrically, . . .

This technique is called the ________ for vector addition because the vector **u** + **v**, often called the ________ of vector addition, is . . .

Let $\mathbf{u} = \langle u_1, u_2 \rangle$ and $\mathbf{v} = \langle v_1, v_2 \rangle$ be vectors and let *k* be a scalar (a real number). **Vector addition,** that is, the sum of **u** and **v,** is defined as the following vector:

u + **v** = ________________

Scalar multiplication, that is, the scalar multiple of *k* times **u,** is defined as the following vector:

*k***u** = ________________

Example 2: Let $\mathbf{u} = \langle 1, 6 \rangle$ and $\mathbf{v} = \langle -4, 2 \rangle$. Find:
(a) 3**u**
(b) **u** + **v**

Let **u**, **v**, and **w** be vectors and c and d be scalars. Complete the following properties of vector addition and scalar multiplication:

1. $\mathbf{u} + \mathbf{v} =$
2. $(\mathbf{u} + \mathbf{v}) + \mathbf{w} =$
3. $\mathbf{u} + \mathbf{0} =$
4. $\mathbf{u} + (-\mathbf{u}) =$
5. $c(d\mathbf{u}) =$
6. $(c + d)\mathbf{u} =$
7. $c(\mathbf{u} + \mathbf{v}) =$
8. $1(\mathbf{u}) =$
9. $0(\mathbf{u}) =$
10. $\|c\mathbf{v}\| =$

IV. Unit Vectors (Pages 448–449)

What you should learn
How to write vectors as linear combinations of unit vectors

To find a unit vector **u** that has the same direction as a given nonzero vector **v**, . . .

In this case, the vector **u** is called a

.

Example 3: Find a unit vector in the direction of $\mathbf{v} = \langle -8, 6 \rangle$.

Let $\mathbf{v} = \langle v_1, v_2 \rangle$. Then the standard unit vectors can be used to represent **v** as **v** = ________________ , where the scalar v_1 is called the and the scalar v_2 is called the . The vector sum $v_1\mathbf{i} + v_2\mathbf{j}$ is called a of the vectors **i** and **j**.

Example 4: Let $\mathbf{v} = \langle -5, 3 \rangle$. Write **v** as a linear combination of the standard unit vectors **i** and **j**.

Example 5: Let $\mathbf{v} = 3\mathbf{i} - 4\mathbf{j}$ and $\mathbf{w} = 2\mathbf{i} + 9\mathbf{j}$. Find $\mathbf{v} + \mathbf{w}$.

V. Direction Angles (Page 450)

> ***What you should learn***
> How to find the direction angles of vectors

If **u** is a unit vector and θ is its direction angle, the terminal point of **u** lies on the unit circle and

$\mathbf{u} = \langle x, y \rangle =$ $\qquad\qquad\qquad$ $=$

Now, if **v** is any vector that makes an angle θ with the positive x-axis, it has the same direction as **u** and

$\mathbf{v} =$ $\qquad\qquad\qquad$ $=$

If **v** can be written as $\mathbf{v} = a\mathbf{i} + b\mathbf{j}$, then the direction angle θ for **v** can be determined from $\tan\theta =$ $\qquad\qquad$.

Example 6: Let $\mathbf{v} = -4\mathbf{i} + 5\mathbf{j}$. Find the direction angle for **v**.

VI. Applications of Vectors (Pages 451–452)

> ***What you should learn***
> How to use vectors to model and solve real-life problems

Describe several real-life applications of vectors.

> **Homework Assignment**
>
> Page(s)
>
> Exercises

Course Number
Instructor
Date

Section 6.4 Vectors and Dot Products

Objective: In this lesson you learned how to find the dot product of two vectors and find the angle between two vectors.

Important Vocabulary Define each term or concept.

Angle between two nonzero vectors

Orthogonal vectors

Vector components

I. The Dot Product of Two Vectors (Pages 458–459)

What you should learn
How to find the dot product of two vectors and use the Properties of the Dot Product

The **dot product** of $\mathbf{u} = \langle u_1, u_2 \rangle$ and $\mathbf{v} = \langle v_1, v_2 \rangle$ is ______________. This product yields a ______________.

Let **u**, **v**, and **w** be vectors in the plane or in space and let c be a scalar. Complete the following properties of the dot product:

1. $\mathbf{u} \bullet \mathbf{v} =$
2. $\mathbf{0} \bullet \mathbf{v} =$
3. $\mathbf{u} \bullet (\mathbf{v} + \mathbf{w}) =$
4. $\mathbf{v} \bullet \mathbf{v} =$
5. $c(\mathbf{u} \bullet \mathbf{v}) =$ $\quad =$

Example 1: Find the dot product: $\langle 5, -4 \rangle \bullet \langle 9, -2 \rangle$.

II. The Angle Between Two Vectors (Pages 459–461)

What you should learn
How to find the angle between two vectors and how to determine whether two vectors are orthogonal

If θ is the angle between two nonzero vectors **u** and **v**, then θ can be determined from ______________.

Example 2: Find the angle between $\mathbf{v} = \langle 5, -4 \rangle$ and $\mathbf{w} = \langle 9, -2 \rangle$.

An alternative way to calculate the dot product between two vectors **u** and **v**, given the angle θ between them, is .

Two vectors **u** and **v** are orthogonal if .

Example 3: Are the vectors $\mathbf{u} = \langle 1, -4 \rangle$ and $\mathbf{v} = \langle 6, 2 \rangle$ orthogonal?

III. Finding Vector Components (Pages 461–462)

> ***What you should learn***
> How to write a vector as the sum of two vector components

Let **u** and **v** be nonzero vectors such that $\mathbf{u} = \mathbf{w}_1 + \mathbf{w}_2$, where $\mathbf{w}_1$ and $\mathbf{w}_2$ are orthogonal and $\mathbf{w}_1$ is parallel to (or a scalar multiple of) **v**. The vectors $\mathbf{w}_1$ and $\mathbf{w}_2$ are called . The vector $\mathbf{w}_1$ is the **projection** of **u** onto **v** and is denoted by ____________________. The vector $\mathbf{w}_2$ is given by ____________________.

Let **u** and **v** be nonzero vectors. The projection of **u** onto **v** is given by $\text{proj}_{\mathbf{v}}\,\mathbf{u} =$

IV. Work (Page 463)

> ***What you should learn***
> How to use vectors to find the work done by a force

The work W done by a constant force **F** as its point of application moves along the vector $\overrightarrow{PQ}$ is given by either of the following:

1.

2.

> **Homework Assignment**
>
> Page(s)
>
> Exercises

Course Number

Instructor

Date

Section 6.5 Trigonometric Form of a Complex Number

Objective: In this lesson you learned how to multiply and divide complex numbers written in trigonometric form and how to find powers and *n*th roots of complex numbers.

Important Vocabulary Define each term or concept.

***n*th roots of unity**

I. The Complex Plane (Page 467)

What you should learn
How to find absolute values of complex numbers

The absolute value of the complex number $a + bi$ is defined as . . .

The absolute value of the complex number $z = a + bi$ is given by $|a + bi| = \sqrt{\qquad}$.

II. Trigonometric Form of a Complex Number (Pages 468–469)

What you should learn
How to write the trigonometric forms of complex numbers

The **trigonometric form of the complex number** $z = a + bi$ is

$z =$,

where $a =$,

$b =$,

$r = \sqrt{\qquad}$, and

$\tan\theta =$.

The number r is the of z, and θ is called an of z.

The trigonometric form of a complex number is also called the .

Larson/Hostetler/Edwards *Precalculus Functions and Graphs: A Graphing Approach, Third Edition*
Larson/Hostetler/Edwards *Precalculus with Limits: A Graphing Approach, Third Edition*
Student Success Organizer

III. Multiplication and Division of Complex Numbers (Pages 469–470)

> ***What you should learn***
> How to multiply and divide complex numbers written in trigonometric form

Let $z_1 = r_1(\cos \theta_1 + i \sin \theta_1)$ and $z_2 = r_2(\cos \theta_2 + i \sin \theta_2)$ be complex numbers. Then:

$z_1 z_2 =$ ______________________________________

$z_1/z_2 =$ ______________________________________

Describe how to find the product of two complex numbers.

Describe how to find the quotient of two complex numbers.

IV. Powers of Complex Numbers (Page 471)

> ***What you should learn***
> How to use De Moivre's Theorem to find powers of complex numbers

State **DeMoivre's Theorem**.

IV. Roots of Complex Numbers (Pages 472–474)

> ***What you should learn***
> How to find *n*th roots of complex numbers

The complex number $u = a + bi$ is an ***n*th root of the complex number** z if .

For a positive integer n, the complex number $z = r(\cos \theta + i \sin \theta)$ has given by $\sqrt[n]{r}\left(\cos \dfrac{\theta + 2\pi k}{n} + i \sin \dfrac{\theta + 2\pi k}{n}\right)$, where $k = 0, 1, 2, \ldots, n - 1$.

Homework Assignment

Page(s)

Exercises

Chapter 7 Systems of Equations and Inequalities

Course Number

Instructor

Date

Section 7.1 Solving Systems of Equations

Objective: In this lesson you learned how to solve a system of equations by substitution and by graphing and how to use systems of equations to model and solve real-life problems.

Important Vocabulary Define each term or concept.

Systems of equations

Solution of a system of equations (in two variables)

Method of substitution

Point of intersection

Break-even point

I. The Method of Substitution (Pages 488–492)

What you should learn
How to use the method of substitution to solve systems of equations in two variables and how to solve systems of equations graphically

To check that the ordered pair (– 3, 4) is the solution of a system of equations, . . .

List the steps necessary for solving a system of equations using the method of substitution.

Explain what is meant by back-substitution.

Example 1: Solve the system of equations using the method of substitution.

$$\begin{cases} 2x + y = 2 \\ x - 2y = -9 \end{cases}$$

To use a graphing utility to solve a system of equations graphically, . . .

Example 2: Solve the system of equations graphically.

$$\begin{cases} x^2 - y = 5 \\ -x + y = -3 \end{cases}$$

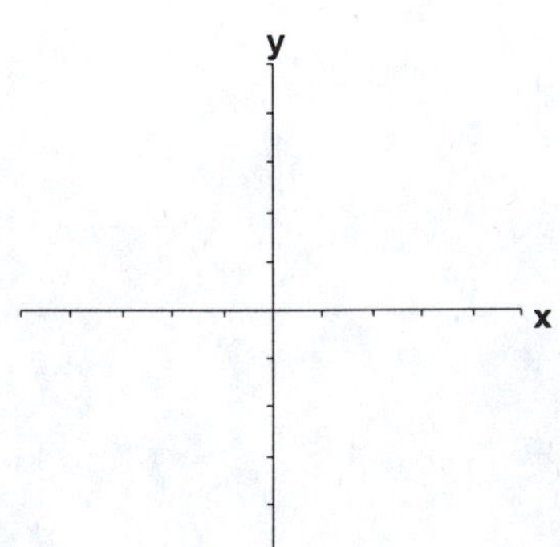

II. Applications of Systems of Equations (Pages 493–494)

What you should learn
How to use systems of equations to model and solve real-life problems

The total cost C of producing x units of a product typically has two components: .

In break-even analysis, the break-even point corresponds to the of the cost and revenue curves.

Break-even analysis can also be approached from the point of view of profit. In this case, consider the profit function, which is . The break-even point occurs when profit equals .

Example 3: The cost of producing x units is $C = 1.5x + 15{,}000$ and the revenue obtained by selling x units is $R = 5x$. How many items should be sold to break even?

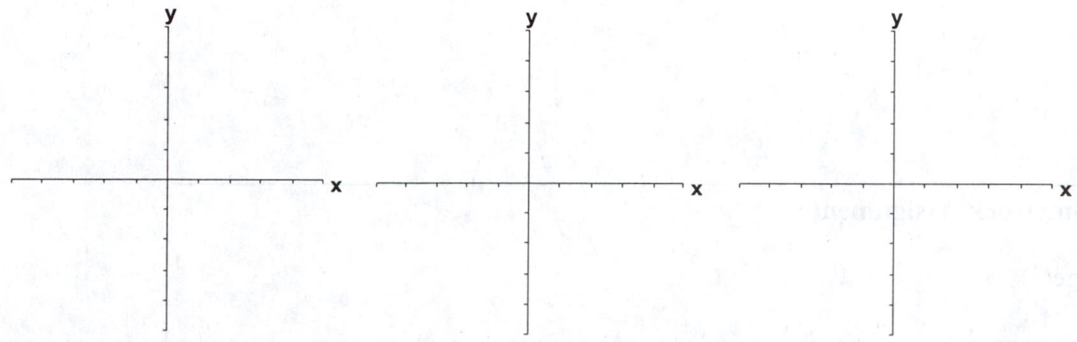

Additional notes

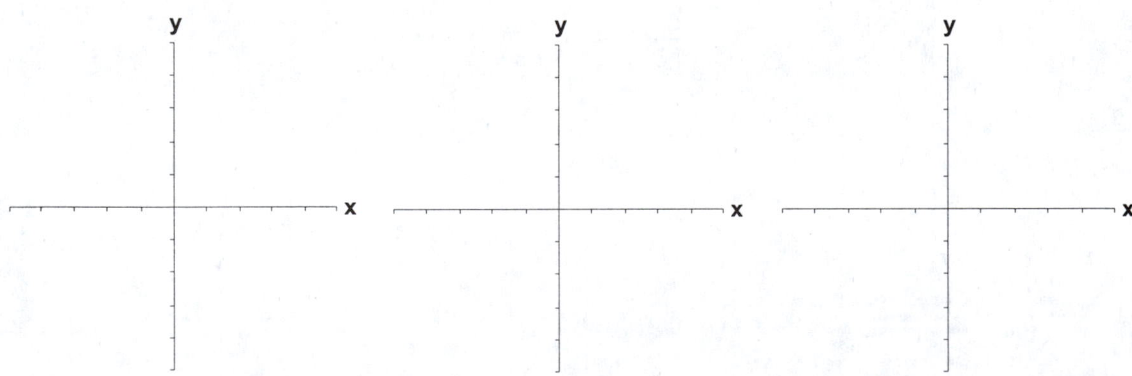

Homework Assignment

Page(s)

Exercises

Section 7.2 Systems of Linear Equations in Two Variables

Course Number

Instructor

Date

Objective: In this lesson you learned how to solve a system of equations by elimination and how to use systems of equations to model and solve real-life problems.

Important Vocabulary Define each term or concept.

Method of elimination

Equivalent systems

Consistent system

Inconsistent system

I. The Method of Elimination (Pages 499–500)

What you should learn
How to use the method of elimination to solve systems of linear equations in two variables

List the steps necessary for solving a system of equations using the method of elimination.

The operations that can be performed on a system of linear equations to produce an equivalent system are:

(1)

(2)

(3)

Example 1: Describe a strategy for solving the system of linear equations using the method of elimination.

$$\begin{cases} 3x + y = 9 \\ 4x - 2y = -1 \end{cases}$$

Example 2: Solve the system of linear equations using the method of elimination.

$$\begin{cases} 4x + y = -3 \\ x - 3y = 9 \end{cases}$$

II. Graphical Interpretation of Two-Variable Systems (Pages 501–503)

> ***What you should learn***
> How to interpret graphically the numbers of solutions of systems of linear equations in two variables

If a system of linear equations has two different solutions, it must have ______________ solutions.

For a system of two linear equations in two variables, list the possible number of solutions the system can have and give a graphical interpretation of the solutions.

If a contradictory statement such as 9 = 0 is obtained while solving a system of linear equations using the method of elimination, then the system has ________ ________.

If a statement that is true for all values of the variable, such as 0 = 0, is obtained while solving a system of linear equations using the method of elimination, then the system has ________________.

Example 3: Is the following system consistent or inconsistent? How many solutions does the system have?

$$\begin{cases} x - 3y = 2 \\ -4x + 12y = 8 \end{cases}$$

III. Applications of Two-Variable Linear Systems (Page 504)

> ***What you should learn***
> How to use systems of equations in two variables to model and solve real-life problems

When may a system of linear equations be an appropriate mathematical model for solving a real-life application?

Give an example of a real-life application that could be solved with a system of linear equations.

Additional notes

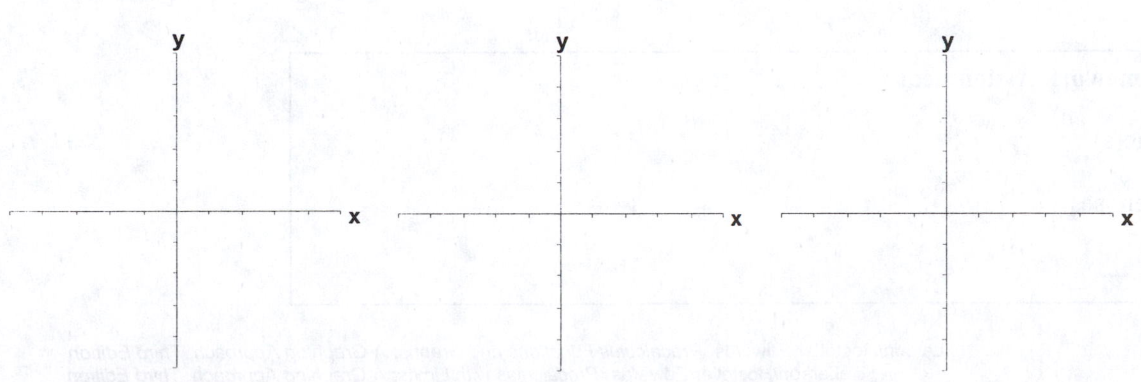

Additional notes

y x y x y x

Homework Assignment

Page(s)

Exercises

Section 7.3 Multivariable Linear Systems

Course Number

Instructor

Date

Objective: In this lesson you learned how to solve a system of equations by Gaussian elimination, how to recognize linear systems in row-echelon form and to use back substitution to solve the system, how to solve nonsquare systems of equations, and how to use a system of equations to model and solve real-life problems.

Important Vocabulary Define each term or concept.

Row-echelon form

Ordered triple

Gaussian elimination

Nonsquare system of equations

Three-dimensional coordinate system

Graph of an equation in three variables

Partial fraction

Partial fraction decomposition

Basic equation

I. Row-Echelon Form and Back-Substitution (Page 509)

What you should learn
How to recognize linear systems in row-echelon form and use back-substitution to solve the systems

When elimination is used to solve a system of linear equations, the goal is . . .

Example 1: Solve the system of linear equations.

$$\begin{cases} x + y - z = 9 \\ \quad\; y - 2z = 4 \\ \qquad\quad z = 1 \end{cases}$$

Larson/Hostetler/Edwards *Precalculus Functions and Graphs: A Graphing Approach, Third Edition*
Larson/Hostetler/Edwards *Precalculus with Limits: A Graphing Approach, Third Edition*
Student Success Organizer

II. Gaussian Elimination (Pages 510–512)

> ***What you should learn***
> How to use Gaussian elimination to solve systems of linear equations

To solve a system that is not in row-echelon form, . . .

List the three elementary row operations that can be used on a system of linear equations to produce an equivalent system of linear equations.

1.

2.

3.

The solution(s) of a system of linear equations in more than two variables must fall into one of the following three categories:

1.

2.

3.

Example 2: Solve the system of linear equations.

$$\begin{cases} x + y + z = 3 \\ 2x - y + 3z = 16 \\ x - 2y - z = 1 \end{cases}$$

A consistent system having exactly one solution is

. A consistent system with infinitely

many solutions is .

Example 3: The following equivalent system is obtained during the course of Gaussian elimination. Write the solution of the system.

$$\begin{cases} x + 2y - z = 4 \\ y + 2z = 8 \\ 0 = 0 \end{cases}$$

III. Nonsquare Systems (Page 513)

> ***What you should learn***
> How to solve nonsquare systems of linear equations

In a square system of linear equations, the number of equations in the system is the number of variables.

If a system has more variables than equations, the system cannot have a(n) .

Example 4: Solve the system of linear equations.

$$\begin{cases} x + y + z = 1 \\ x - 2y - 2z = 4 \end{cases}$$

IV. Graphical Interpretation of Three-Variable Systems (Page 514)

> ***What you should learn***
> How to graphically interpret three-variable systems

To sketch the graph of a plane, . . .

The graph of a system of three linear equations in three variables consists of planes. When these planes intersect in a single point, the system has solution(s). When the planes have no point in common, the system has solution(s). When the planes intersect in a line or a plane, the system has solution(s).

V. Partial Fraction Decomposition and Other Applications

> ***What you should learn***
> How to use systems of linear equations to write partial fraction decompositions of rational expressions and to use systems of linear equations in three or more variables to model and solve real-life problems

Suppose the rational expression $N(x)/D(x)$ is an improper fraction. Before the expression can be decomposed into partial fractions, you must . . .

To decompose a proper rational expression into partial fractions, completely factor the denominator into factors of the form

and , where

is irreducible.

Describe how to deal with both linear factors and quadratic factors in the next step of a partial fraction decomposition.

To find the **basic equation** of a partial fraction decomposition, . . .

To solve the basic equation, . . .

Example 5: Write the form of the partial fraction decomposition for $\dfrac{x-4}{x^2-8x+12}$.

Example 6: Solve the basic equation $5x+3=A(x-1)+B(x+3)$ for A and B.

Homework Assignment
Page(s)
Exercises

Course Number

Instructor

Date

Section 7.4 Systems of Inequalities

Objective: In this lesson you learned how to sketch the graphs of inequalities in two variables and solve systems of inequalities and how to use systems of inequalities to model and solve real-life problems.

Important Vocabulary Define each term or concept.

Solution of an inequality

Graph of an inequality

Linear inequalities

Solution of a system of inequalities

I. The Graph of an Inequality (Pages 525–526)

What you should learn
How to sketch the graphs of inequalities in two variables

To sketch the graph of an inequality in two variables, . . .

The solution points for the inequality $y < 3x + 5$ lie the line $y = 3x + 5$.

Example 1: Sketch the graph of the linear inequality $y \geq 2$.

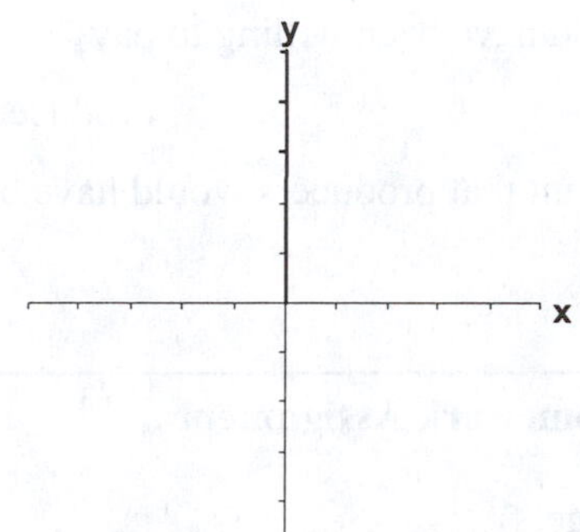

II. Systems of Inequalities (Pages 527–529)

> ***What you should learn***
> How to solve systems of inequalities

To sketch the graph of a system of inequalities in two variables, . . .

To find the vertices of the solution region for a system of three linear inequalities, . . .

III. Applications of Systems of Inequalities (Pages 530–531)

> ***What you should learn***
> How to use systems of inequalities in two variables to model and solve real-life problems

The   is defined by the price p and the number of units x that satisfy both the demand and supply equations.

Consumer surplus is defined as . . .

Producer surplus is defined as . . .

The consumer surplus is a measure of the amount that consumers would have been willing to pay

. Producer surplus is a measure of the amount that producers would have been willing to receive

.

Homework Assignment

Page(s)

Exercises

Course Number
Instructor
Date

Section 7.5 Linear Programming

Objective: In this lesson you learned how to solve linear programming problems.

Important Vocabulary	Define each term or concept.
Optimization	
Linear programming	
Objective function	
Constraints	
Feasible solutions	

I. Linear Programming: A Graphical Approach (Pages 535–538)

What you should learn
How to solve linear programming problems

If a linear programming problem has a solution, it must occur . . .

If there is more than one solution to a linear programming problem, at least one of them . . .

In either case, the value of the objective function is .

List the steps for solving a linear programming problem:

Example 1: The vertices of the region of feasible solutions for a linear programming problem are as follows:
(0, 0)
(5, 0)
(10, 3)
(7, 6)
(0, 4)
If the objective function is $z = 8x + 3y$, find the maximum value and where it occurs.

Example 2: Find the minimum value of $z = 4x + 6y$ subject to the following constraints.

$$\left.\begin{array}{l} x \geq 0 \\ y \geq 0 \\ x + y \geq 2 \\ y \leq 4 \\ x \leq 5 \end{array}\right\}$$

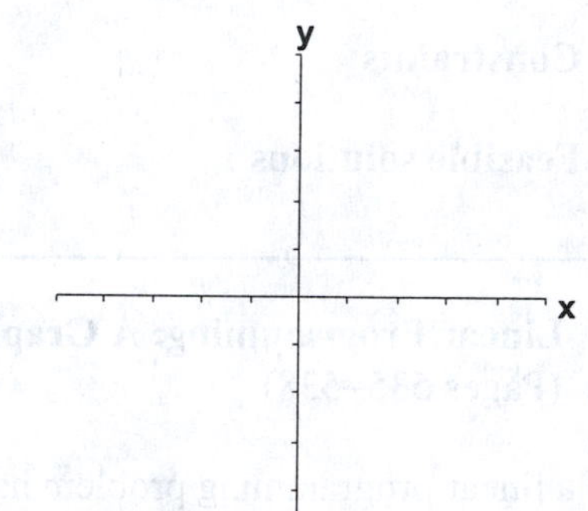

II. Applications of Linear Programming (Pages 539–540)

Describe a real-life problem that can be solved using linear programming.

What you should learn
How to use linear programming to model and solve real-life problems

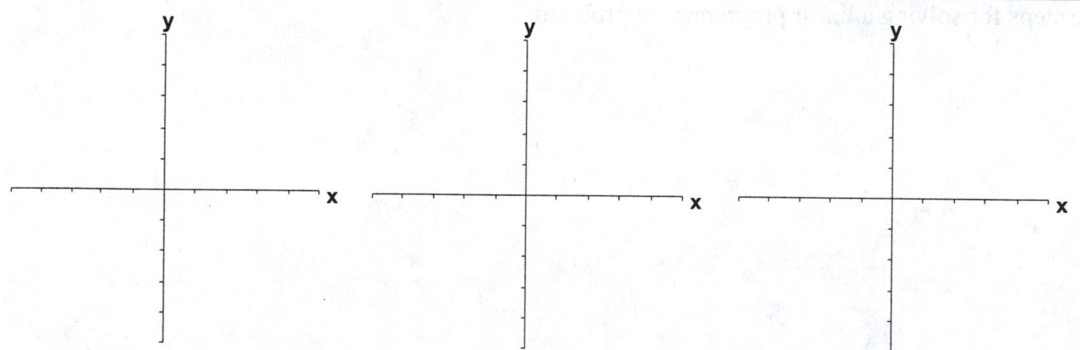

Homework Assignment

Page(s)

Exercises

Chapter 8 Matrices and Determinants

Course Number
Instructor
Date

Section 8.1 Matrices and Systems of Equations

Objective: In this lesson you learned how to write matrices, identify their order, and perform elementary row operations and how to use Gaussian elimination and Gauss-Jordan elimination with matrices to solve systems of linear equations.

Important Vocabulary Define each term or concept.

Entry of a matrix

Order of a matrix

Square matrix

Main diagonal

Row matrix

Column matrix

Elementary row operations

Gauss-Jordan elimination

I. Matrices (Pages 552–553)

What you should learn
How to write matrices and identify their orders

If m and n are positive integers, an $m \times n$ **matrix** is . . .

An $m \times n$ matrix has rows and columns.

An **augmented matrix** is . . .

A **coefficient matrix** is . . .

Example 1: Consider the following system of equations.

$$\begin{cases} 2x + y - z = 5 \\ x - 3y + 2z = 9 \\ 3x + 2y = 1 \end{cases}$$

(a) Write the augmented matrix for this system.
(b) What is the order of the augmented matrix?
(c) Write the coefficient matrix for this system.
(d) What is the order of the coefficient matrix?

II. Elementary Row Operations (Page 554)

What you should learn
How to perform elementary row operations on matrices

The **elementary row operations** on a matrix are:

Two matrices are **row-equivalent** if . . .

III. Gaussian Elimination with Back-Substitution (Pages 555–558)

What you should learn
How to use matrices and Gaussian elimination to solve systems of linear equations

A matrix in **row-echelon form** has the following three properties:
1.

2.

3.

A matrix in row-echelon form is in **reduced row-echelon form** if . . .

To solve a system of linear equations using Gaussian Elimination with Back-Substitution, . . .

If, during the elimination process, you obtain a row with zeros except for the last entry, you can conclude that the system has

.

Example 2: Solve the following system using Gaussian Elimination with Back-Substitution.

$$\begin{cases} x+y+z=1 \\ x+2y+3z=1 \\ x-3y+5z=-11 \end{cases}$$

IV. Gauss-Jordan Elimination (Pages 559–561)

What you should learn
How to use matrices and Gauss-Jordan elimination to solve systems of linear equations

Example 3: Apply Gauss-Jordan elimination to the following matrix to obtain the unique reduced row-echelon form of the matrix.

$$\left[\begin{array}{ccc:c} 1 & 4 & 2 & 5 \\ 0 & 1 & -1 & 3 \\ 0 & 0 & 1 & -2 \end{array}\right]$$

Example 4: Solve the following system using Gauss-Jordan elimination.

$$\begin{cases} 2x - y + 3z = 1 \\ x + 2y - 4z = -6 \\ -2x + 3y - z = 13 \end{cases}$$

Homework Assignment

Page(s)

Exercises

Course Number
Instructor
Date

Section 8.2 Operations with Matrices

Objective: In this lesson you learned how to add, subtract, and multiply two matrices, and multiply a matrix by a real number.

Important Vocabulary Define each term or concept.

Scalar multiple

Zero matrix

Additive identity

Matrix multiplication

Identity matrix of order *n*

I. Equality of Matrices (Page 567)

What you should learn
How to decide whether two matrices are equal

Name three ways that a matrix may be represented.

1)

2)

3)

Two matrices are equal if they have the same order and

are equal.

II. Matrix Addition and Scalar Multiplication (Pages 568–571)

What you should learn
How to add and subtract matrices and multiply matrices by real numbers

To add two matrices of the same order, . . .

To multiply a matrix A by a scalar c, . . .

Example 1: Let $A = \begin{bmatrix} 2 & 5 \\ -3 & 1 \end{bmatrix}$ and $B = \begin{bmatrix} -1 & 4 \\ 2 & -5 \end{bmatrix}$.

Find (a) $A + B$ and (b) $-2B$

Let A, B, and C be $m \times n$ matrices and let c and d be scalars. Give an example of each of the following properties of matrix addition and scalar multiplication:

1) Commutative Property of Matrix Addition:

2) Associative Property of Matrix Addition:

3) Associative Property of Scalar Multiplication:

4) Scalar Identity:

5) Distributive Property (two forms):

If A is an $m \times n$ matrix and O is the $m \times n$ zero matrix, then

$A + O =$

III. Matrix Multiplication (Pages 572–574)

What you should learn
How to multiply two matrices

When multiplying an $m \times n$ matrix A by an $n \times p$ matrix B, to obtain the entry in the ith row and jth column of AB, . . .

.

Example 2: If A is a 3×5 matrix and B is a 6×3 matrix, find the order, if possible, of the product (a) AB, and (b) BA.

Example 3: Find the product AB, if

$$A = \begin{bmatrix} 2 & -1 & 7 \\ 0 & 6 & -3 \end{bmatrix} \quad \text{and} \quad B = \begin{bmatrix} 0 \\ -2 \\ 3 \end{bmatrix}$$

List four properties of Matrix Multiplication:

If A is an $n \times n$ matrix, the identity matrix I of order n has the property that and .

IV. Applications of Matrix Operations (Pages 575–576)

What you should learn
How to use matrix operations to model and solve real-life problems

Matrix multiplication can be used to represent a system of linear equations. The system

$$\begin{cases} a_{11}x_1 + a_{12}x_2 + a_{13}x_3 = b_1 \\ a_{21}x_1 + a_{22}x_2 + a_{23}x_3 = b_2 \\ a_{31}x_1 + a_{32}x_2 + a_{33}x_3 = b_3 \end{cases}$$

can be written as the matrix equation ,
where A is the coefficient matrix of the system and X and B are column matrices.

Example 4: Consider the following system of linear equations.

$$\begin{cases} 2x_1 - x_2 + 3x_3 = -11 \\ x_1 - 3x_3 = -1 \\ -x_1 + 4x_2 + 2x_3 = 2 \end{cases}$$

Write this system as a matrix equation $AX = B$, and then use Gauss-Jordan elimination on the augmented matrix $[A : B]$ to solve for the matrix X.

Additional notes

Homework Assignment

Page(s)

Exercises

Course Number

Instructor

Date

Section 8.3 The Inverse of a Square Matrix

Objective: In this lesson you learned how to verify that two matrices are inverses of each other and find inverses of matrices and how to use inverse matrices to solve systems of linear equations.

Important Vocabulary Define each term or concept.

Inverse of a matrix

I. The Inverse of a Matrix (Pages 582–583)

What you should learn
How to verify that two matrices are inverses of each other

To verify that a matrix B is the inverse of the matrix A, . . .

If a matrix A has an inverse, A is called or **nonsingular.** Otherwise, A is called .

To have an inverse, a matrix must be . Not all square matrices have inverses. However, if a matrix does have an inverse, that inverse is .

II. Finding Inverse Matrices (Pages 584–585)

What you should learn
How to use Gauss-Jordan elimination to find the inverses of matrices

To find the inverse of a square matrix A of order n, . . .

Example 1: Find the inverse of the matrix $A = \begin{bmatrix} 1 & 2 & 4 \\ 1 & 0 & 2 \\ 2 & 3 & 6 \end{bmatrix}$.

III. The Inverse of a 2 × 2 Matrix (Page 586)

> ***What you should learn***
> How to use a formula to find the inverses of 2 × 2 matrices

If A is a 2 × 2 matrix given by $A = \begin{bmatrix} a & b \\ c & d \end{bmatrix}$, then A is invertible if and only if ____________ . Moreover, if this condition is true, the inverse of A is given by:

$$A^{-1} = \frac{\qquad}{\qquad}\begin{bmatrix} \quad & \quad \\ \quad & \quad \end{bmatrix}$$

The denominator is called the ____________ of the 2 × 2 matrix A.

Example 2: Find the inverse of the matrix $B = \begin{bmatrix} 3 & 9 \\ -2 & -7 \end{bmatrix}$.

IV. Systems of Linear Equations (Page 587)

> ***What you should learn***
> How to use inverse matrices to solve systems of linear equations

If A is an invertible matrix, the system of linear equations represented by $AX = B$ has a unique solution given by ____________ .

Example 3: Use an inverse matrix to solve (if possible) the system of linear equations:

$$\begin{cases} 12x + 8y = 416 \\ 3x + 5y = 152 \end{cases}$$

Homework Assignment
Page(s)
Exercises

Course Number

Instructor

Date

Section 8.4 The Determinant of a Square Matrix

Objective: In this lesson you learned how to find determinants of square matrices.

Important Vocabulary Define each term or concept.

Determinant

Minors

Cofactors

I. The Determinant of a Matrix (Pages 591–592)

What you should learn
How to find the determinants of 2×2 matrices

The **determinant** of the 2×2 matrix $A = \begin{bmatrix} a_1 & b_1 \\ a_2 & b_2 \end{bmatrix}$ is given by

$$\det(A) = |A| = \left| \quad \right| = \underline{\qquad\qquad}$$

The determinant of a matrix of order 1×1 is defined as . . .

Example 1: Find the determinant of the matrix $A = \begin{bmatrix} -4 & 3 \\ 1 & -2 \end{bmatrix}$.

II. Minors and Cofactors (Page 593)

What you should learn
How to find minors and cofactors of square matrices

Complete the sign patterns for cofactors of a 3×3 matrix, a 4×4 matrix, and a 5×5 matrix:

Sign Pattern for Cofactors

3×3 matrix $\qquad$ 4×4 matrix $\qquad$ 5×5 matrix

Example 2: Use the matrix $A = \begin{bmatrix} 1 & 0 & 3 \\ 2 & 1 & 0 \\ 0 & 2 & 3 \end{bmatrix}$ to find:

(a) the minor M_{13}, and (b) the cofactor C_{21}.

III. The Determinant of a Square Matrix (Page 594)

What you should learn
How to find the determinants of square matrices

Applying the definition of the determinant of a square matrix to find a determinant is called ____________________.

Example 3: Find the determinant of the matrix:

$$A = \begin{bmatrix} -1 & 0 & 4 \\ 3 & -2 & 0 \\ 1 & -1 & 1 \end{bmatrix}$$

Example 4: Describe a strategy for finding the determinant of the following matrix, and then find the determinant of the matrix.

$$B = \begin{bmatrix} -2 & 4 & 0 & 5 \\ 0 & 2 & -1 & 0 \\ 3 & 1 & -4 & -1 \\ -5 & 0 & -2 & 3 \end{bmatrix}$$

IV. The Determinant of a Square Matrix (Page 595)

What you should learn
How to find the determinants of square matrices

A **triangular matrix** is . . .

A square matrix is   if it has all zero entries below its main diagonal and is   if it has all zero entries above its main diagonal.

A **diagonal matrix** is . . .

To find the determinant of a triangular matrix, . . .

Example 5: Find the determinant of the following matrix:

$$A = \begin{bmatrix} 3 & -1 & 2 & 5 & -6 & -2 \\ 0 & -1 & 3 & -4 & 2 & 1 \\ 0 & 0 & 2 & -2 & -2 & 5 \\ 0 & 0 & 0 & 1 & -3 & -1 \\ 0 & 0 & 0 & 0 & 4 & 8 \\ 0 & 0 & 0 & 0 & 0 & -2 \end{bmatrix}$$

Additional notes

Homework Assignment

Page(s)

Exercises

Course Number

Instructor

Date

Section 8.5 Applications of Matrices and Determinants

Objective: In this lesson you learned how to use Cramer's Rule to solve systems of linear equations.

Important Vocabulary Define each term or concept.

Uncoded row matrices

Coded row matrices

I. Area of a Triangle (Page 599)

What you should learn
How to use determinants to find areas of triangles

The area of a triangle with vertices (x_1, y_1), (x_2, y_2), and (x_3, y_3) is

$$\text{Area} = \pm\frac{1}{2}\left|\quad\quad\right|$$

where the symbol ± indicates that the appropriate sign should be chosen to yield a positive area.

Example 1: Find the area of a triangle whose vertices are (– 3, 1), (2, 4), and (5, – 3).

II. Collinear Points (Page 600)

What you should learn
How to use determinants to decide whether points are collinear

Collinear points are . . .

Three points (x_1, y_1), (x_2, y_2), and (x_3, y_3) are collinear if and only if

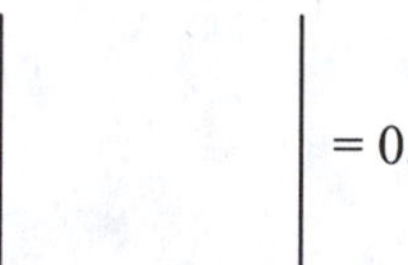

Example 2: Determine whether the points (– 2, 4), (0, 3), and (8, – 1) are collinear.

Larson/Hostetler/Edwards *Precalculus Functions and Graphs: A Graphing Approach, Third Edition*
Larson/Hostetler/Edwards *Precalculus with Limits: A Graphing Approach, Third Edition*
Student Success Organizer

III. Cramer's Rule (Pages 601–603)

What you should learn
How to use Cramer's Rule to solve systems of linear equations

Cramer's Rule states that if a system of n linear equations in n variables has a coefficient matrix A with a nonzero determinant $|A|$, the solution of the system is

$$x_1 = \frac{|A_1|}{|A|}, \quad x_2 = \frac{|A_2|}{|A|}, \ldots, x_n = \frac{|A_n|}{|A|}$$

where the ith column of A_i is

.

Cramer's Rule does not apply if the determinant of the coefficient matrix is , in which case the system has either no solution or .

Example 3: Use Cramer's Rule to solve the system of linear equations.

$$\begin{cases} 2x + y + z = 6 \\ -x - y + 3z = 1 \\ y - 2z = -3 \end{cases}$$

IV. Cryptography (Pages 604–606)

What you should learn
How to use matrices to code and decode messages

A cryptogram is . . .

To use matrix multiplication to encode and decode messages, . . .

Homework Assignment

Page(s)

Exercises

Chapter 9 Sequences, Series, and Probability

Course Number
Instructor
Date

Section 9.1 Sequences and Series

Objective: In this lesson you learned how to use sequence, factorial, and summation notation to write the terms and sums of sequences.

Important Vocabulary	Define each term or concept.
Recursive	

I. Sequences (Pages 618–620)

What you should learn
How to use sequence notation to write the terms of a sequence

An **infinite sequence** is . . .

The function values $a_1, a_2, a_3, a_4, \ldots, a_n, \ldots$ are the

of an infinite sequence.

A **finite sequence** is . . .

To find the first three terms of a sequence, given an expression for its *n*th term, . . .

Example 1: Find the first five terms of the sequence given by $a_n = 5 + 2n(-1)^n$.

II. Factorial Notation (Pages 620–621)

What you should learn
How to use factorial notation

If *n* is a positive integer, ***n* factorial** is defined by

By definition, zero factorial is .

Example 2: Evaluate the factorial expression $\frac{n!}{(n+1)!}$.

III. Summation Notation (Pages 622–623)

> ***What you should learn***
> How to use summation notation to write sums

The sum of the first n terms of a sequence is represented by the **summation or sigma notation,**

$$\sum_{i=1}^{n} a_i = \underline{\qquad\qquad\qquad\qquad}$$

where i is called the , n is the , and 1 is the .

Example 3: Find the following sum: $\sum_{i=2}^{7}(2+3i)$.

IV. Series (Page 623)

> ***What you should learn***
> How to find the sum of an infinite series

The sum of the terms of a finite or infinite sequence is called a .

Consider the infinite sequence $a_1, a_2, a_3, \ldots, a_i, \ldots$. The sum of all terms of the infinite sequence is called a(n)

and is denoted by $a_1 + a_2 + a_3 + \cdots + a_i + \cdots = \sum_{i=1}^{\infty} a_i$. The sum of

the first n terms of the sequence is called a(n)

or the of the sequence and is denoted by

$$a_1 + a_2 + a_3 + \cdots + a_n = \sum_{i=1}^{n} a_i.$$

Homework Assignment

Page(s)

Exercises

Course Number

Instructor

Date

Section 9.2 Arithmetic Sequences and Partial Sums

Objective: In this lesson you learned how to recognize, write, and use arithmetic sequences.

Important Vocabulary Define each term or concept.

Arithmetic sequence

I. Arithmetic Sequences (Pages 629–631)

What you should learn
How to recognize and write arithmetic sequences

The common difference of an arithmetic sequence is . . .

The *n*th term of an arithmetic sequence has the form ____________________, where d is the common difference between consecutive terms of the sequence, and $c = a_1 - d$. Therefore, an arithmetic sequence may be thought of as a(n) ____________ function whose domain is the set of natural numbers.

Example 1: Determine whether or not the following sequence is arithmetic. If it is, find the common difference.
7, 3, – 1, – 5, – 9, . . .

Example 2: Find a formula for the *n*th term of the arithmetic sequence whose common difference is 2 and whose first term is 7.

The *n*th term of an arithmetic sequence has the alternative recursive formula ______________________________.

Example 3: Find the sixth term of the arithmetic sequence that begins with 15 and 12.

II. The Sum of a Finite Arithmetic Sequence (Pages 632–633)

> ***What you should learn***
> How to find an *n*th partial sum of an arithmetic sequence

The sum of a finite arithmetic sequence with *n* terms is

____.

The sum of the first *n* terms of an infinite sequence is the

____.

Example 4: Find the sum of the first 20 terms of the sequence with *n*th term $a_n = 28 - 5n$.

III. Applications of Arithmetic Sequences (Pages 633–634)

> ***What you should learn***
> How to use arithmetic sequences to model and solve real-life problems

Describe a real-life problem that could be solved by finding the sum of a finite arithmetic sequence.

Additional notes

Homework Assignment

Page(s)

Exercises

Course Number
Instructor
Date

Section 9.3 Geometric Sequences and Series

Objective: In this lesson you learned how to recognize, write, and use geometric sequences.

Important Vocabulary Define each term or concept.

Geometric sequence

Infinite geometric series or geometric series

I. Geometric Sequences (Pages 638–640)

What you should learn
How to recognize and write geometric sequences

The common ratio of a geometric sequence is . . .

The *n*th term of a geometric sequence has the form

__________________, where r is the common ratio of consecutive terms of the sequence. So, every geometric sequence can be written in the following form:

____.

A geometric sequence may be thought of as a(n) function whose domain is the set of natural numbers.

Example 1: Determine whether or not the following sequence is geometric. If it is, find the common ratio.
60, 30, 0, – 30, – 60, . . .

To find the $(n + 1)$th term of a geometric sequence given the nth term of the same sequence, . . .

Example 2: Write the first five terms of the geometric sequence whose first term is $a_1 = 5$ and whose common ratio is – 3.

Example 3: Find the eighth term of the geometric sequence that begins with 15 and 12.

II. The Sum of a Finite Geometric Sequence (Page 641)

The sum of the geometric sequence $a_1, a_1r, a_1r^2, a_1r^3, a_1r^4, \ldots,$ a_1r^{n-1} with common ratio $r \neq 1$ is given by

____.

When using the formula for the sum of a geometric sequence, be careful to check that the index begins with $i = 1$. If the index begins at $i = 0, \ldots$

> ***What you should learn***
> How to find an *n*th partial sum of a geometric sequence

Example 4: Find the sum $\sum_{i=1}^{10} 2(0.5)^i$.

III. Geometric Series (Page 642)

If $|r| < 1$, the sum of the infinite geometric series $a_1, a_1r, a_1r^2,$ $a_1r^3, a_1r^4, \ldots, a_1r^{n-1}, \ldots$ is ____________________________.

> ***What you should learn***
> How to find the sum of an infinite geometric series

Example 5: If possible, find the sum: $\sum_{i=1}^{\infty} 9(0.25)^{i-1}$.

IV. Applications of Geometric Sequences (Page 643)

Describe a real-life problem that could be solved by finding the sum of a finite geometric sequence.

> ***What you should learn***
> How to use geometric sequences to model and solve real-life problems

> **Homework Assignment**
>
> Page(s)
>
> Exercises

Course Number

Instructor

Date

Section 9.4 Mathematical Induction

Objective: In this lesson you learned how to use mathematical induction to prove statements involving a positive integer *n*.

Important Vocabulary Define each term or concept.

Mathematical induction

I. Introduction (Pages 648–651)

What you should learn
How to use mathematical induction to prove a statement

To apply the Principle of Mathematical Induction, you need to be able to determine the statement ________________ for a given statement P_k.

When using mathematical induction to prove a summation formula, it is helpful to think of S_{k+1} as . . .

Describe the process needed to prove a formula using mathematical induction.

The extended principle of mathematical induction is . . .

II. Sums of Powers of Integers (Page 652)

> ***What you should learn***
> How to find the sums of powers of integers

List the formulas for the following sums of powers of integers.

1. $1 + 2 + 3 + 4 + \cdots + n =$

2. $1^2 + 2^2 + 3^2 + 4^2 + \cdots + n^2 =$

3. $1^3 + 2^3 + 3^3 + 4^3 + \cdots + n^3 =$

4. $1^4 + 2^4 + 3^4 + 4^4 + \cdots + n^4 =$

5. $1^5 + 2^5 + 3^5 + 4^5 + \cdots + n^5 =$

III. Finite Differences (Page 653)

> ***What you should learn***
> How to find finite differences of a sequence

First differences are . . .

When the first differences of a sequence are all the same, the sequence has a model.

Second differences are . . .

When the second differences of a sequence are all the same, the sequence has a model.

Homework Assignment

Page(s)

Exercises

Course Number

Instructor

Date

Section 9.5 The Binomial Theorem

Objective: In this lesson you learned how to use the Binomial Theorem and Pascal's Triangle to calculate binomial coefficients and write binomial expansions.

Important Vocabulary Define each term or concept.

Binomial coefficients

Pascal's Triangle

I. Binomial Coefficients (Pages 656–657)

What you should learn
How to use the Binomial Theorem to calculate binomial coefficients

List four general observations about the expansion of $(x + y)^n$ for various values of n.

1)

2)

3)

4)

The **Binomial Theorem** states that in the expansion of $(x + y)^n =$ $x^n + nx^{n-1}y + \ldots + {}_nC_r x^{n-r}y^r + \ldots + nxy^{n-1} + y^n$, the coefficient of $x^{n-r}y^r$ is ______________________.

Example 1: Find the binomial coefficient ${}_{12}C_5$.

II. Pascal's Triangle (Page 658)

What you should learn
How to use Pascal's Triangle to calculate binomial coefficients

Construct rows 0 through 6 of Pascal's Triangle.

III. Binomial Expansions (Pages 659–660)

What you should learn
How to use binomial coefficients to write binomial expansions

Writing out the coefficients for a binomial that is raised to a power is called ______________.

Example 2: Use the binomial coefficients from the appropriate row of Pascal's Triangle to expand $(x + 2)^5$

Additional notes

Homework Assignment

Page(s)

Exercises

Course Number

Instructor

Date

Section 9.6 Counting Principles

Objective: In this lesson you learned how to solve counting problems using the Fundamental Counting Principle, permutations, and combinations.

Important Vocabulary Define each term or concept.

Fundamental Counting Principle

Permutation

Distinguishable permutations

I. Simple Counting Problems (Page 664)

What you should learn
How to solve simple counting problems

If two balls are randomly drawn from a bag of six balls, numbered from 1 to 6, such that it is possible to choose two 3's, the random selection occurs _________________________. If two balls are drawn from the bag at the same time, the random selection occurs _________________________, which eliminates the possibility of choosing two 3's.

II. The Fundamental Counting Principle (Page 665)

What you should learn
How to use the Fundamental Counting Principle to solve counting problems

The Fundamental Counting Principle can be extended to three or more events. For instance, if E_1 can occur in m_1 ways, E_2 in m_2 ways, and E_3 in m_3 ways, the number of ways that three events E_1, E_2, and E_3 can occur is _________________________.

Example 1: A diner offers breakfast combination plates which can be made from a choice of one of 4 different types of breakfast meats, one of 8 different styles of eggs, and one of 5 different types of breakfast breads. How many different breakfast combination plates are possible?

III. Permutations (Pages 666–668)

What you should learn
How to use permutations to solve counting problems

The number of different ways that n elements can be ordered is .

A **permutation of *n* elements taken *r* at a time** is . . .

Example 2: In how many ways can a chairperson, a vice chairperson, and a recording secretary be chosen from a committee of 14 people?

Example 3: In how many distinguishable ways can the letters COMMITTEE be written?

IV. Combinations (Pages 669–670)

What you should learn
How to use combinations to solve counting problems

A **combination of *n* elements taken *r* at a time** is . . .

Example 4: In how many ways can a research team of 3 students be chosen from a class of 14 students?

Homework Assignment

Page(s)

Exercises

Course Number

Instructor

Date

Section 9.7 Probability

Objective: In this lesson you learned how to find the probability of events and their complements.

Important Vocabulary Define each term or concept.

Independent events

Complement of an event

I. The Probability of an Event (Pages 674–677)

What you should learn
How to find the probability of an event

An happening whose result is uncertain is called a(n) . The possible results of the experiment are , the set of all possible outcomes of the experiment is the of the experiment, and any subcollection of a sample space is a(n) .

The measure of the likelihood that an event will occur based on chance is called the of an event. If an event E has $n(E)$ equally likely outcomes and its sample space S has $n(S)$ equally likely outcomes, the probability of event E is .

The probability of an event must be between and .

If $P(E) = 0$, the event E occur, and E is called a(n) event. If $P(E) = 1$, the event E occur, and E is called a(n) event.

Example 1: A box contains 3 red marbles, 5 black marbles, and 2 yellow marbles. If a marble is selected at random from the box, what is the probability that it is yellow?

II. Mutually Exclusive Events (Pages 678–679)

What you should learn
How to find the probabilities of mutually exclusive events

Two events *A* and *B* (from the same sample space) are ________ if *A* and *B* have no outcomes in common.

If *A* and *B* are events in the same sample space, the probability of *A or B* occurring is given by $P(A \cup B) =$ ________.

To find the probability that one or the other of two mutually exclusive events will occur, . . .

Example 2: A box contains 3 red marbles, 5 black marbles, and 2 yellow marbles. If a marble is selected at random from the box, what is the probability that it is either red or black?

III. Independent Events (Page 680)

What you should learn
How to find the probabilities of independent events

If *A* and *B* are independent events, the probability that both *A* and *B* will occur is $P(A \text{ and } B) =$ ________.

That is, to find the probability that two independent events will occur, . . .

Example 3: A box contains 3 red marbles, 5 black marbles, and 2 yellow marbles. If two marbles are randomly selected with replacement, what is the probability that both marbles are yellow?

IV. The Complement of an Event (Page 681)

What you should learn
How to find the probability of the complement of an event

Let *A* be an event and let *A′* be its complement. If the probability of *A* is $P(A)$, the probability of the complement is $P(A') =$ ________.

Homework Assignment
Page(s)

Exercises

Chapter 10 Topics in Analytic Geometry

Section 10.1 Introduction to Conics: Parabolas

Course Number

Instructor

Date

Objective: In this lesson you learned how to write the standard equation of a parabola, and analyze and sketch the graphs of parabolas.

Important Vocabulary Define each term or concept.

Directrix

Focus

Tangent

I. Conics (Page 696)

What you should learn
How to recognize a conic as the intersection of a plane and a double-napped cone

A **conic section,** or **conic,** is . . .

Name the four basic conic sections:

In the formation of the four basic conics, the intersecting plane does not pass through the vertex of the cone. When the plane does pass through the vertex, the resulting figure is a(n)

, such as . . .

II. Parabolas (Pages 697–699)

What you should learn
How to write equations of parabolas in standard form

A **parabola** is . . .

The of a parabola is the midpoint between the focus and the directrix. The of the parabola is the line passing through the focus and the vertex.

The standard form of the equation of a parabola with a vertical axis having a vertex at (h, k) and directrix $y = k - p$ is

____

The standard form of the equation of a parabola with a horizontal axis having a vertex at (h, k) and directrix $x = h - p$ is

____

The focus lies on the axis p units (directed distance) from the vertex. If the vertex is at the origin $(0, 0)$, the equation takes on one of the following forms:

Example 1: Find the standard form of the equation of the parabola with vertex at the origin and focus $(1, 0)$.

III. Applications of Parabolas (Pages 699–700)

What you should learn
How to use the reflective property of parabolas to solve real-life problems

Describe a real-life situation in which parabolas are used.

A focal chord is . . .

The specific focal chord perpendicular to the axis of a parabola is called the

The reflective property of a parabola states that the tangent line to a parabola at a point P makes equal angles with the following two lines:

1)

2)

Homework Assignment

Page(s)

Exercises

Course Number
Instructor
Date

Section 10.2 Ellipses

Objective: In this lesson you learned how to write the standard equation of an ellipse, and analyze and sketch the graphs of ellipses.

Important Vocabulary	Define each term or concept.
Foci	
Vertices	
Major axis	
Center	
Minor axis	

I. Introduction (Pages 704–707)

What you should learn How to write equations of ellipses in standard form

An **ellipse** is . . .

The standard form of the equation of an ellipse centered at (h, k) and having a horizontal major axis of length $2a$ and minor axis of length $2b$, where $0 < b < a$, is: ____________________

The standard form of the equation of an ellipse centered at (h, k) and having a vertical major axis of length $2a$ and minor axis of length $2b$, where $0 < b < a$, is: ____________________

In both cases, the foci lie on the major axis, c units from the center, with $c^2 =$.

If the center is at the origin (0, 0), the equation takes one of the following forms: or .

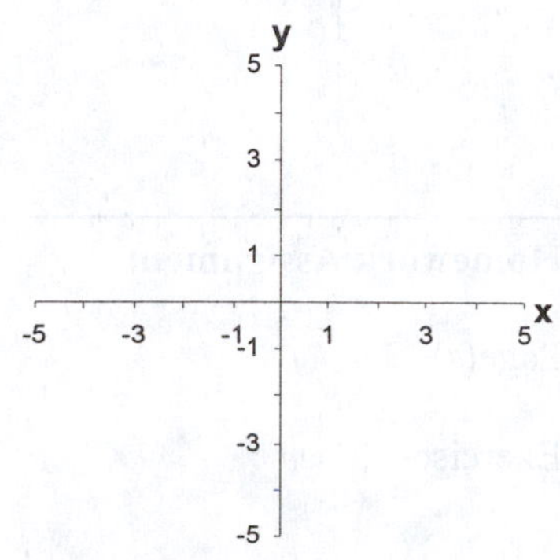

Example 1: Sketch the ellipse given by $4x^2 + 25y^2 = 100$.

II. Applications of Ellipses (Page 708)

Describe a real-life application in which parabolas are used.

> ***What you should learn***
> How to use properties of ellipses to model and solve real-life problems

III. Eccentricity (Pages 708–709)

> ***What you should learn***
> How to find the eccentricities of ellipses

measures the ovalness of an ellipse. It is given by the ratio $e =$. For every ellipse, the value of e lies between and . For an elongated ellipse, the value of e is close to

Additional notes

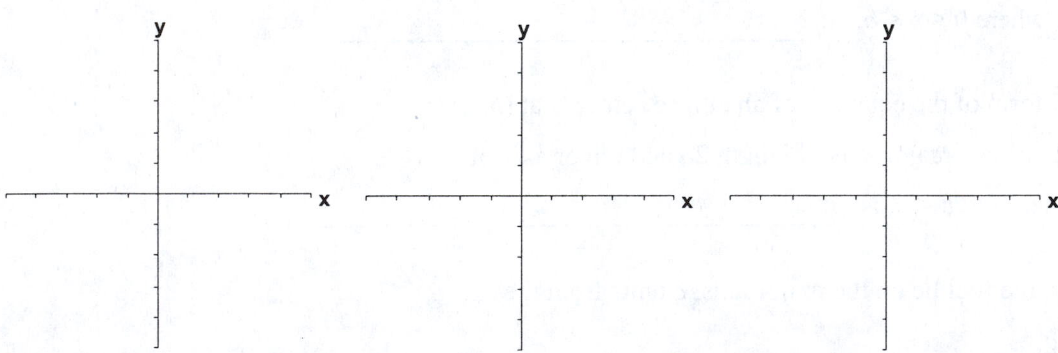

Homework Assignment

Page(s)

Exercises

Course Number

Instructor

Date

Section 10.3 Hyperbolas

Objective: In this lesson you learned how to write the standard equation of a hyperbola, and analyze and sketch the graphs of hyperbolas.

Important Vocabulary Define each term or concept.

Branches

Transverse axis

Conjugate axis

I. Introduction (Pages 713–714)

What you should learn
How to write equations of hyperbolas in standard form

A **hyperbola** is . . .

The line through a hyperbola's two foci intersects the hyperbola at two points called .

The midpoint of a hyperbola's transverse axis is the of the hyperbola.

The standard form of the equation of a hyperbola centered at (h, k) and having a horizontal transverse axis is

____

The standard form of the equation of a hyperbola centered at (h, k) and having a vertical transverse axis is

____

In each case, the vertices and foci are, respectively, a and c units from the center. Moreover, a, b, and c are related by the equation .

If the center of the hyperbola is at the origin $(0, 0)$, the equation takes one of the following forms: or .

Larson/Hostetler/Edwards *Precalculus Functions and Graphs: A Graphing Approach, Third Edition*
Larson/Hostetler/Edwards *Precalculus with Limits: A Graphing Approach, Third Edition*
Student Success Organizer

II. Asymptotes of a Hyperbola (Pages 715–717)

What you should learn
How to find asymptotes of hyperbolas

The **asymptotes** of a hyperbola with a horizontal transverse axis are .

The **asymptotes** of a hyperbola with a vertical transverse axis are .

Example 1: Sketch the graph of the hyperbola given by $y^2 - 9x^2 = 9$.

The **eccentricity** of a hyperbola is $e =$, where the values of e are .

III. Applications of Hyperbolas (Page 718–719)

What you should learn
How to use properties of hyperbolas to solve real-life problems and to classify conics from their general equations

Describe a real-life application in which hyperbolas occur or are used.

The graph of $Ax^2 + Cy^2 + Dx + Ey + F = 0$ is one of the following:

1) Circle if

2) Parabola if

3) Ellipse if

4) Hyperbola if

Example 2: Classify the equation $9x^2 + y^2 - 18x - 4y + 4 = 0$ as a circle, a parabola, an ellipse, or a hyperbola.

Homework Assignment

Page(s)

Exercises

Course Number

Instructor

Date

Section 10.4 Rotation of Conics

Objective: In this lesson you learned how to rotate the coordinate axis to eliminate the *xy*-term in equations of conics, use the discriminant to classify conics, and solve systems of quadratic equations.

Important Vocabulary Define each term or concept.

Discriminant

I. Rotation (Pages 722–725)

What you should learn
How to rotate the coordinate axes to eliminate the *xy*-term in equations of conics

The general equation of a conic whose axes are rotated so that they are not parallel to either the *x*-axis or the *y*-axis contains a(n) .

To eliminate this term, you can use a procedure called , whose goal is to rotate the *x*- and *y*-axes until they are parallel to the axes of the conic.

The general second-degree equation

$Ax^2 + Bxy + Cy^2 + Dx + Ey + F = 0$ can be rewritten as

$A'(x')^2 + C'(y')^2 + D'x' + E'y' + F' = 0$ by rotating the coordinate axes through an angle θ, where

$\cot 2\theta =$.

The coefficients of the new equation are obtained by making the substitutions $x =$ and

$y =$.

II. Invariants Under Rotation (Pages 726–727)

What you should learn
How to use the discriminant to classify conics

Invariant under rotation means . . .

The rotation of the coordinate axes through an angle θ that transforms the equation $Ax^2 + Bxy + Cy^2 + Dx + Ey + F = 0$ into the form $A'(x')^2 + C'(y')^2 + D'x' + E'y' + F' = 0$ has the following rotation invariants:

1)

2)

3)

The graph of the equation $Ax^2 + Bxy + Cy^2 + Dx + Ey + F = 0$ is, except in degenerate cases, determined by its discriminant as follows:

1) Ellipse or circle if:

2) Parabola if:

3) Hyperbola if:

Example 1: Classify the graph of the following conic:
$2x^2 + 12xy + 18y^2 - 3y - 5 = 0$

III. Systems of Quadratic Equations (Page 728)

What you should learn
How to solve systems of quadratic equations

To find the points of intersection of two conics, . . .

Example 2: Solve the following system of quadratic equations:

$$\begin{cases} 4x^2 + 4y^2 - 36 = 0 \\ x^2 - 3y - 6x + 9 = 0 \end{cases}$$

Homework Assignment

Page(s)

Exercises

Course Number

Instructor

Date

Section 10.5 Parametric Equations

Objective: In this lesson you learned how to rewrite sets of parametric equations as single rectangular equations and find sets of parametric equations for graphs.

Important Vocabulary Define each term or concept.

Parameter

I. Plane Curves (Page 731)

What you should learn
How to evaluate sets of parametric equations for given values of the parameter

If f and g are continuous functions of t on an interval I, the set of ordered pairs $(f(t), g(t))$ is a(n) ________ C. The equations $x = f(t)$ and $y = g(t)$ are ________ for C, and t is the ________.

II. Sketching a Plane Curve (Pages 732–733)

What you should learn
How to graph curves that are represented by sets of parametric equations

One way to sketch a curve represented by a pair of parametric equations is to plot points in the ________. Each set of coordinates (x, y) is determined from a value chosen for the ________. By plotting the resulting points in the order of increasing values of t, you trace the curve in a specific direction, called the ________ of the curve.

Example 1: Sketch the curve described by the parametric equations $x = t - 3$ and $y = t^2 + 1$, $-1 \le t \le 3$.

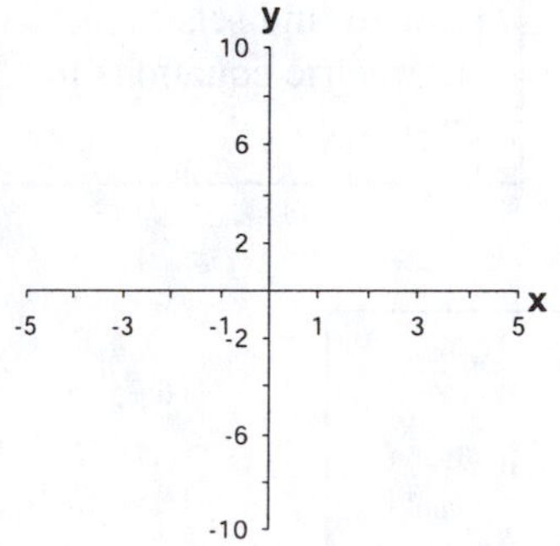

Another way to display a curve represented by a pair of parametric equations is to use a graphing utility. To do so, . . .

III. Eliminating the Parameter (Pages 734–735)

What you should learn
How to rewrite sets of parametric equations as single rectangular equations by eliminating the parameter

Eliminating the parameter is the process of . . .

Describe the process used to eliminate the parameter from a set of parametric equations.

Converting equations from parametric to rectangular form can change the of x and y. In such cases, . . .

To eliminate the parameter in equations involving trigonometric functions, try using the identities . . .

IV. Finding Parametric Equations for a Graph (Page 735)

What you should learn
How to find sets of parametric equations for graphs

Describe how to find a set of parametric equations for a given graph.

Homework Assignment

Page(s)

Exercises

Course Number

Instructor

Date

Section 10.6 Polar Coordinates

Objective: In this lesson you learned how to plot points in the polar coordinate system and write equations in polar form.

I. Introduction (Pages 739–740)

What you should learn
How to plot points and find multiple representations of points in the polar coordinate system

To form the **polar coordinate system** in the plane, fix a point O, called the ______ or ______, and construct from O an initial ray called the ______. Then each point P in the plane can be assigned ______ as follows:

1) $r =$

2) $\theta =$

In the polar coordinate system, points do not have a unique representation. For instance, the point (r, θ) can be represented as ______ or ______, where n is any integer. Moreover, the pole is represented by $(0, \theta)$, where θ is

Example 1: Plot the point $(r, \theta) = (-2, 11\pi/4)$ on the polar coordinate system.

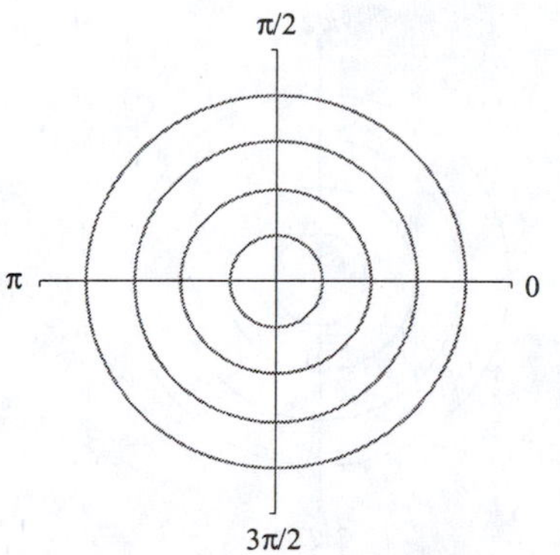

Example 2: Find another polar representation of the point $(4, \pi/6)$.

II. Coordinate Conversion (Pages 740–741)

What you should learn
How to convert points from rectangular to polar form and vice versa

The polar coordinates (r, θ) are related to the rectangular coordinates (x, y) as follows . . .

Example 3: Convert the polar coordinates $(3, 3\pi/2)$ to rectangular coordinates.

III. Equation Conversion (Page 742)

What you should learn
How to convert equations from rectangular to polar form and vice versa

To convert a rectangular equation to polar form, . . .

Example 4: Find the rectangular equation corresponding to the polar equation $r = \dfrac{-5}{\sin\theta}$.

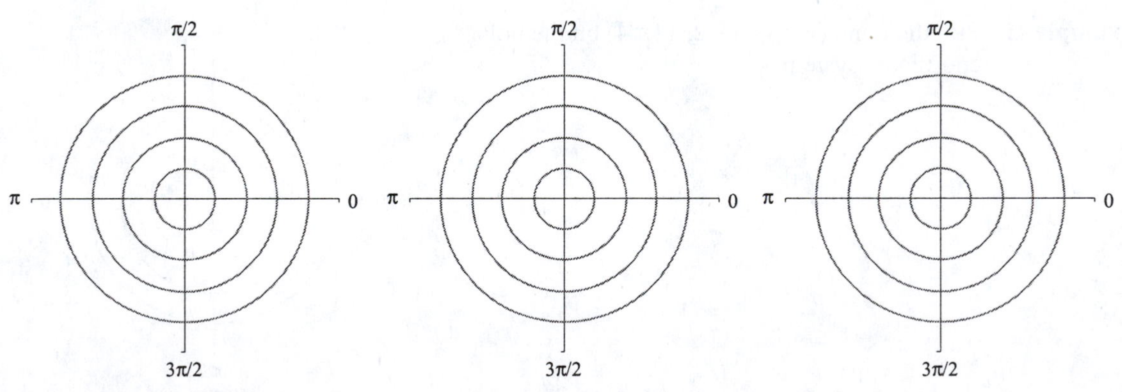

Homework Assignment

Page(s)

Exercises

Course Number

Instructor

Date

Section 10.7 Graphs of Polar Equations

Objective: In this lesson you learned how to graph polar equations and recognize special polar graphs.

I. Introduction (Pages 745–746)

What you should learn How to graph polar equations by point plotting

Example 1: Use point plotting to sketch the graph of the polar equation $r = 3\cos\theta$.

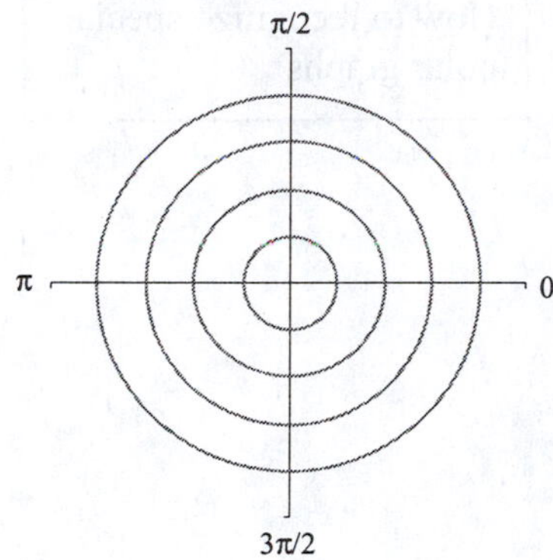

The graph of the polar equation $r = f(\theta)$ can be written in parametric form, using t as a parameter, as follows:

II. Symmetry (Pages 746–747)

What you should learn How to use symmetry as an aid to graphing polar equations

The graph of a polar equation is symmetric with respect to the following if the given substitution yields an equivalent equation.

Substitution

1) The line $\theta = \pi/2$:

2) The polar axis:

3) The pole:

Example 2: Describe the symmetry of the polar equation $r = 2(1 - \sin\theta)$.

III. Zeros and Maximum r-Values (Pages 748–749)

What you should learn How to use zeros and maximum r-values as graphing aids

Two additional aids to sketching graphs of polar equations are . . .

Example 3: Describe the zeros and maximum r-values of the polar equation $r = 5\cos 2\theta$

IV. Special Polar Graphs (Pages 750–751)

What you should learn
How to recognize special polar graphs

List the general equations that yield each of the following types of special polar graphs:

Limaçons:

Rose curves:

Circles:

Lemniscates:

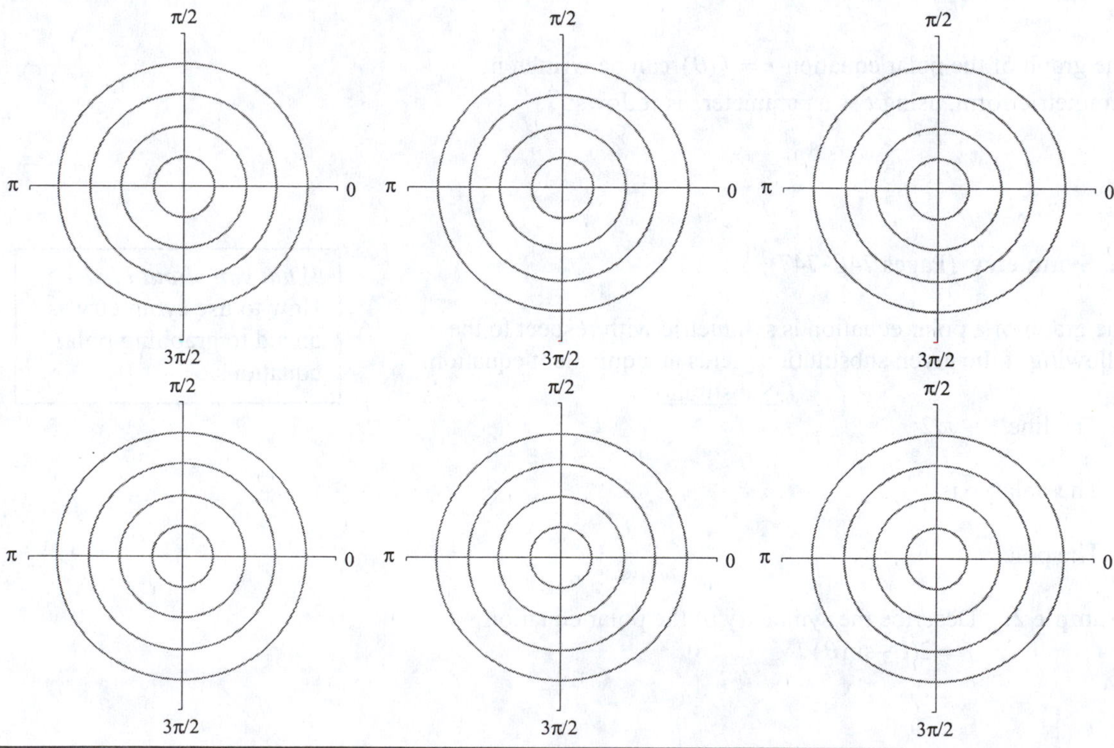

Homework Assignment

Page(s)

Exercises

Course Number

Instructor

Date

Section 10.8 Polar Equations of Conics

Objective: In this lesson you learned how to define a conic in terms of eccentricity and to write equations of conics in polar form.

I. Alternative Definition of Conics (Page 754)

What you should learn
How to define conics in terms of eccentricity

The locus of a point in the plane that moves so that its distance from a fixed point (focus) is in a constant ratio to its distance from a fixed line (directrix) is a ________. The constant ratio is the ________ of the conic and is denoted by e. Moreover, the conic is an ellipse if ________, a parabola if ________, and a hyperbola if ________.

For each type of conic, the ________ of the polar coordinate system corresponds to the fixed point (focus) given in the above definition.

II. Polar Equations of Conics (Pages 754–756)

What you should learn
How to write equations of conics in polar form

The graph of the polar equation ________ is a conic with a vertical directrix to the right of the pole, where $e > 0$ is the eccentricity and $|p|$ is the distance between the focus (pole) and the directrix.

The graph of the polar equation ________ is a conic with a vertical directrix to the left of the pole, where $e > 0$ is the eccentricity and $|p|$ is the distance between the focus (pole) and the directrix.

The graph of the polar equation ________ is a conic with a horizontal directrix above the pole, where $e > 0$ is the eccentricity and $|p|$ is the distance between the focus (pole) and the directrix.

The graph of the polar equation

is a conic with a horizontal directrix below the pole, where $e > 0$ is the eccentricity and $|p|$ is the distance between the focus (pole) and the directrix.

Example 1: Identify the type of conic from the polar equation $r = \dfrac{36}{10 + 12\sin\theta}$, and describe its orientation.

III. Applications (Page 757)

Describe a real-life application of polar equations of conics.

> ***What you should learn***
> How to use equations of conics in polar form to model real-life problems

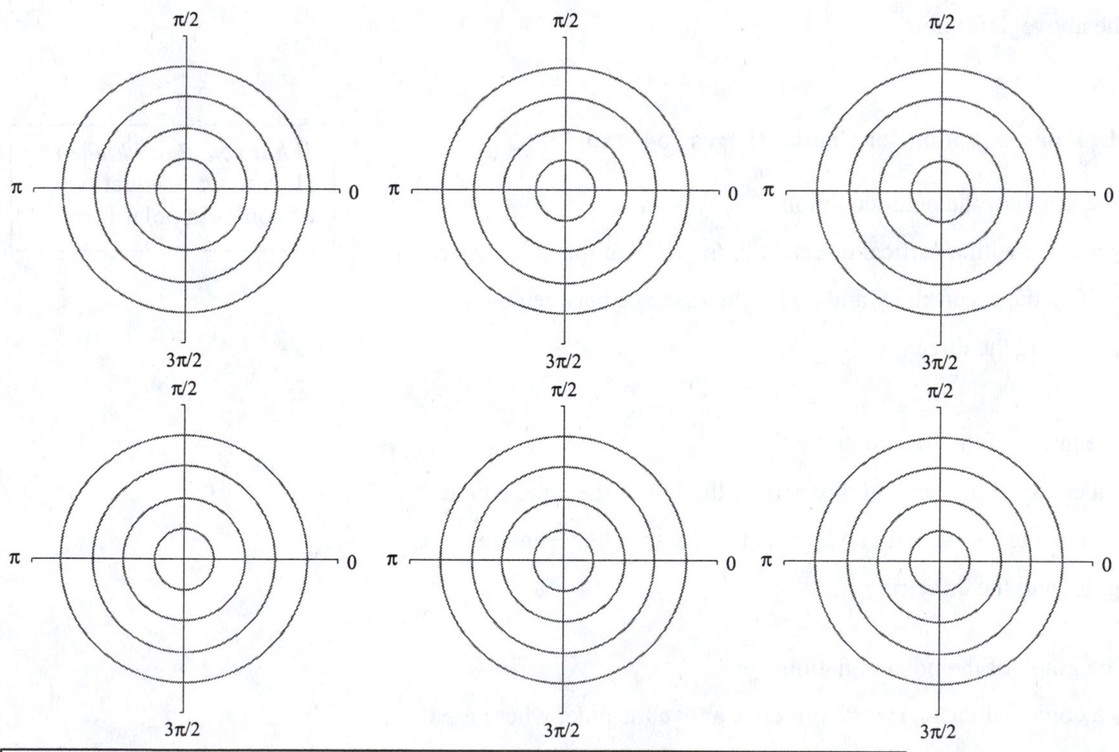

Homework Assignment

Page(s)

Exercises

Chapter 11 Analytic Geometry in Three Dimensions

Course Number

Instructor

Date

Section 11.1 The Three-Dimensional Coordinate System

Objective: In this lesson you learned how to plot points, find distances between points, and find midpoints of line segments connecting points in space and how to write equations of spheres and graph traces of surfaces in space.

Important Vocabulary Define each term or concept.

Solid analytic geometry

Sphere

Surface in space

Trace

I. The Three-Dimensional Coordinate System (Pages 770–771)

What you should learn
How to plot points in the three-dimensional coordinate system

A **three-dimensional coordinate system** is constructed by . . .

Taken as pairs, the axes determine three coordinate planes: the ____________, the ____________, and the ____________.

These three coordinate planes separate the three-dimensional coordinate system into eight ____________. The first of these is the one for which . . .

In the three-dimensional system, a point P in space is determined by an ordered triple (x, y, z), where x, y, and z are as follows . . .

II. The Distance and Midpoint Formulas (Pages 771–772)

What you should learn
How to find distances between points in space and midpoints of line segments joining points in space

The distance between the points (x_1, y_1, z_1) and (x_2, y_2, z_2) given by the **Distance Formula in Space** is

The midpoint of the line segment joining the points (x_1, y_1, z_1) and (x_2, y_2, z_2) given by the **Midpoint Formula in Space** is

$\left(\qquad \right)$

Example 1: For the points (2, 0, – 4) and (– 1, 4, 6), find
(a) the distance between the two points, and
(b) the midpoint of the line segment joining them.

III. The Equations of a Sphere (Pages 772–774)

What you should learn
How to write equations of spheres in standard form and how to find traces of surfaces in space

The standard equation of a sphere whose center is (h, k, j) and whose radius is r is .

Example 2: Find the center and radius of the sphere whose equation is $x^2 + y^2 + z^2 - 4x + 2y + 8z + 17 = 0$.

To find the *yz*-trace of a surface, . . .

Homework Assignment

Page(s)

Exercises

Course Number
Instructor
Date

Section 11.2 Vectors in Space

Objective: In this lesson you learned how to represent vectors and find dot products of and angles between vectors in space.

Important Vocabulary Define each term or concept.

Standard unit vector notation in space

Angle between two nonzero vectors in space

Parallel vectors in space

I. Vectors in Space (Pages 777–779)

What you should learn
How to find the component form, the unit vector in the same direction, and magnitude of vectors in space; as well as how to find dot products of and angles between vectors in space

In space, vectors are denoted by ordered triples of the form

________________________________ .

The **zero vector in space** is denoted by .

If **v** is represented by the directed line segment from $P(p_1, p_2, p_3)$ to $Q(q_1, q_2, q_3)$, the **component form** of **v** is produced by . . .

Two vectors are equal if and only if . . .

The magnitude(or length) of $\mathbf{u} = \langle u_1, u_2, u_3 \rangle$ is:

$\| \mathbf{u} \| = \sqrt{\qquad}$ ________________

A unit vector **u** in the direction of **v** is .

The sum of $\mathbf{u} = \langle u_1, u_2, u_3 \rangle$ and $\mathbf{v} = \langle v_1, v_2, v_3 \rangle$ is

$\mathbf{u} + \mathbf{v} =$ ________________________________ .

The scalar multiple of the real number c and $\mathbf{u} = \langle u_1, u_2, u_3 \rangle$ is

$c\mathbf{u} =$ ______________________.

The dot product of $\mathbf{u} = \langle u_1, u_2, u_3 \rangle$ and $\mathbf{v} = \langle v_1, v_2, v_3 \rangle$ is

$\mathbf{u} \bullet \mathbf{v} =$ ______________________

If θ is the angle between two nonzero vectors **u** and **v**, then θ can be determined from   .

If the dot product of two nonzero vectors is zero, the angle between the vectors is   . Such vectors are called   .

Example 1: Find the dot product of the vectors $\langle -1, 4, -2 \rangle$ and $\langle 0, -1, 5 \rangle$.

II. Parallel Vectors (Pages 779–780)

> ***What you should learn***
> How to determine whether vectors in space are parallel or orthogonal

Example 2: Determine whether the vectors $\langle 6, 1, -3 \rangle$ and $\langle -2, -1/3, 1 \rangle$ are parallel.

To use vectors to determine whether three points P, Q, and R in space are collinear, . . .

III. Applications of Vectors in Space (Page 781)

> ***What you should learn***
> How to use vectors in space to solve real-life problems

Describe a real-life application of vectors in space.

Homework Assignment

Page(s)

Exercises

Course Number

Instructor

Date

Section 11.3 The Cross Product of Two Vectors

Objective: In this lesson you learned how to find cross products of vectors in space, use geometric properties of the cross product, and use triple scalar products to find volumes of parallelepipeds.

I. The Cross Product (Pages 784–785)

What you should learn
How to find cross products of vectors in space

A vector in space that is orthogonal to two given vectors is called their ________________ .

Let $\mathbf{u} = u_1\mathbf{i} + u_2\mathbf{j} + u_3\mathbf{k}$ and $\mathbf{v} = v_1\mathbf{i} + v_2\mathbf{j} + v_3\mathbf{k}$ be two vectors in space. The cross product of **u** and **v** is the vector

$\mathbf{u} \times \mathbf{v} =$ ________________________________

Describe a convenient way to remember the formula for the cross product.

Example 1: Given $\mathbf{u} = -2\mathbf{i} + 3\mathbf{j} - 3\mathbf{k}$ and $\mathbf{v} = \mathbf{i} - 2\mathbf{j} + \mathbf{k}$, find the cross product $\mathbf{u} \times \mathbf{v}$.

Let **u**, **v**, and **w** be vectors in space and let c be a scalar. Complete the following properties of the cross product:

1. $\mathbf{u} \times \mathbf{v} =$
2. $\mathbf{u} \times (\mathbf{v} + \mathbf{w}) =$
3. $c(\mathbf{u} \times \mathbf{v}) =$
4. $\mathbf{u} \times \mathbf{0} =$
5. $\mathbf{u} \times \mathbf{u} =$
6. $\mathbf{u} \bullet (\mathbf{v} \times \mathbf{w}) =$

II. Geometric Properties of the Cross Product (Pages 786–787)

> ***What you should learn***
> How to use geometric properties of cross products of vectors in space

Complete the following geometric properties of the cross product, given **u** and **v** are nonzero vectors in space and θ is the angle between **u** and **v**.

1. $\mathbf{u} \times \mathbf{v}$ is orthogonal to ______________________.
2. $\|\mathbf{u} \times \mathbf{v}\| =$ ______________________.
3. $\mathbf{u} \times \mathbf{v} = \mathbf{0}$ if and only if ______________________.
4. $\|\mathbf{u} \times \mathbf{v}\| =$ area of the parallelogram having ______________________.

III. The Triple Scalar Product (Page 788)

> ***What you should learn***
> How to use triple scalar products to find volumes of parallelepipeds

For vectors **u**, **v**, and **w** in space, the dot product of **u** and $\mathbf{v} \times \mathbf{w}$ is called the ______________________ of **u**, **v**, and **w**, and is found as

$$\mathbf{u} \bullet (\mathbf{v} \times \mathbf{w}) = \left| \quad \right|$$

The volume V of a parallelepiped with vectors **u**, **v**, and **w** as adjacent edges is ______________________.

Example 2: Find the volume of the parallelepiped having $\mathbf{u} = 2\mathbf{i} + \mathbf{j} - 3\mathbf{k}$, $\mathbf{v} = \mathbf{i} - 2\mathbf{j} + 3\mathbf{k}$, and $\mathbf{w} = 4\mathbf{i} - 3\mathbf{k}$ as adjacent edges.

Homework Assignment

Page(s)

Exercises

Course Number

Instructor

Date

Section 11.4 Lines and Planes in Space

Objective: In this lesson you learned how to find parametric and symmetric equations of lines in space and find distances between points and planes in space.

I. Lines in Space (Pages 791–792)

What you should learn
How to find parametric and symmetric equations of lines in space

The **direction vector v** for the line L through the point $P = (x_1, y_1, z_1)$ is the vector $\mathbf{v} = \langle a, b, c \rangle$ to which . The values a, b, and c are the .

One way of describing the line L is . . .

A line L parallel to the vector $\mathbf{v} = \langle a, b, c \rangle$ and passing through the point $P = (x_1, y_1, z_1)$ is represented by the following parametric equations, where t is the parameter:

If the direction numbers a, b, and c are all nonzero, you can eliminate the parameter t to obtain the **symmetric equations** of a line:

II. Planes in Space (Pages 793–795)

What you should learn
How to find equations of planes in space

The plane containing the point (x_1, y_1, z_1) and having normal vector $\mathbf{n} = \langle a, b, c \rangle$ can be represented by the **standard form of the equation of a plane,** which is

____

By regrouping terms, you obtain the **general form of the equation of a plane** in space:

To find a normal vector to a plane given the general form of the equation of the plane, . . .

Two distinct planes in three-space either are

or

If two distinct planes intersect, the **angle θ between the two planes** is equal to the angle between vectors $\mathbf{n}_1$ and $\mathbf{n}_2$ normal to the two intersecting planes, and is given by

Consequently, two planes with normal vectors $\mathbf{n}_1$ and $\mathbf{n}_2$ are

1. ______________ if $\mathbf{n}_1 \bullet \mathbf{n}_2 = 0$.

2. if $\mathbf{n}_1$ is a scalar multiple of $\mathbf{n}_2$.

III. Sketching Planes in Space (Page 796)

What you should learn
How to sketch planes in space

If a plane in space intersects one of the coordinate planes, the line of intersection is called the of the given plane in the coordinate plane.

To sketch a plane in space, . . .

The plane with equation $3y - 2z + 1 = 0$ is parallel to

IV. Distance Between a Point and a Plane (Page 797)

What you should learn
How to find distances between points and planes in space

The **distance between a plane and a point** Q (not in the plane) is

____

where P is a point in the plane and $\mathbf{n}$ is normal to the plane.

Homework Assignment

Page(s)

Exercises

Chapter 12 Limits and an Introduction to Calculus

Course Number
Instructor
Date

Section 12.1 Introduction to Limits

Objective: In this lesson you learned how to estimate limits and use properties and operations of limits.

I. The Limit Concept and Definition of Limit (Pages 806–808)

What you should learn
How to use the definition of a limit to estimate limits

Define **limit.**

Describe how to estimate the limit $\lim_{x \to -2} \frac{x^2 + 4x + 4}{x + 2}$ numerically.

The existence or nonexistence of $f(x)$ when $x = c$ has no bearing on the existence of . . .

II. Limits That Fail to Exist (Pages 809–810)

What you should learn
How to decide whether limits of functions exist

The limit of $f(x)$ as $x \to c$ does not exist if any of the following conditions is true:

1.

2.

3.

Give an example of a limit that does not exist.

Larson/Hostetler/Edwards *Precalculus Functions and Graphs: A Graphing Approach, Third Edition*
Larson/Hostetler/Edwards *Precalculus with Limits: A Graphing Approach, Third Edition*
Student Success Organizer

III. Properties of Limits (Pages 811–812)

> ***What you should learn***
> How to use properties and operations of limits to find limits

Let b and c be real numbers and let n be a positive integer. Complete each of the following properties of limits.

1. $\lim_{x \to c} b =$

2. $\lim_{x \to c} x =$

3. $\lim_{x \to c} x^n =$

4. $\lim_{x \to c} \sqrt[n]{x} =$

Let b and c be real numbers, let n be a positive integer, and let f and g be functions with the following limits.

$$\lim_{x \to c} f(x) = L \quad \text{and} \quad \lim_{x \to c} g(x) = K$$

Complete each of the following statements about operations with limits.

1. Scalar multiple: $\lim_{x \to c}[b\,f(x)] =$

2. Sum or difference: $\lim_{x \to c}[f(x) \pm g(x)] =$

3. Product: $\lim_{x \to c}[f(x) \cdot g(x)] =$

4. Quotient: $\lim_{x \to c} \dfrac{f(x)}{g(x)} =$

5. Power: $\lim_{x \to c}[f(x)]^n =$

Example 1: Find the limit: $\lim_{x \to 2} \dfrac{4 - x^2}{x}$.

> **Homework Assignment**
>
> Page(s)
>
> Exercises

Course Number

Instructor

Date

Section 12.2 Techniques for Evaluating Limits

Objective: In this lesson you learned how to find limits by direct substitution and by using the dividing out and rationalizing techniques.

I. Limits of Polynomial and Rational Functions (Page 816)

What you should learn
How to find limits of polynomial and rational functions by direct substitution

If p is a polynomial function and c is a real number, then

$\lim_{x \to c} p(x) =$

If r is a rational function given by $r(x) = p(x)/q(x)$, and c is a real number such that $q(c) \neq 0$, then

$\lim_{x \to c} r(x) =$

II. Dividing Out Technique (Pages 817–818)

What you should learn
How to use the dividing out technique to find limits of functions

The validity of the dividing out technique stems from . . .

The dividing out technique should be applied only when . . .

An **indeterminate form** is . . .

When you encounter an indeterminate form by direct substitution into a rational function, you can conclude . . .

Example 1: Find the following limit: $\lim_{x \to 3} \dfrac{x^2 - 8x + 15}{x - 3}$.

III. Rationalizing Technique (Page 819)

What you should learn
How to use the rationalizing technique to find limits of functions

Another way to find the limit of a function is to rationalize the numerator of the function when an indeterminate form is obtained. This is called the ______________________________, which is based on multiplication by a convenient form of 1.

IV. Using Technology (Page 820)

What you should learn
How to approximate limits of functions graphically and numerically

To find limits of nonalgebraic functions, . . .

V. One-Sided Limits (Pages 821–822)

What you should learn
How to evaluate one-sided limits of functions

A **one-sided limit** is . . .

Existence of a Limit
If f is a function and c and L are real numbers, then $\lim_{x \to c} f(x) = L$

if and only if . . .

Homework Assignment

Page(s)

Exercises

Course Number
Instructor
Date

Section 12.3 The Tangent Line Problem

Objective: In this lesson you learned how to approximate slopes of tangent lines, use the limit definition of slope, and use derivatives to find slopes of graphs.

I. Tangent Line to a Graph (Page 826)

What you should learn
How to define the tangent line to a graph

The **tangent line** to a graph of a function f at a point $P(x_1, y_1)$ is

. . .

To determine the rate at which a graph rises or falls at a single point, . . .

II. Slope of a Graph (Page 827)

What you should learn
How to use a tangent line to approximate the slope of a graph at a point

To visually approximate the slope of a graph at a point, . . .

III. Slope and the Limit Process (Pages 828–830)

What you should learn
How to use the limit definition of slope to find exact slopes of graphs

A **secant line** to a graph is . . .

A **difference quotient** is . . .

Give the definition of the slope of a graph.

Example 1: Use the limit process to find the slope of the graph of $f(x) = x^2 + 5$ at the point (3, − 1).

IV. The Derivative of a Function (Pages 831–832)

What you should learn
How to find derivatives of functions and use derivatives to find slopes of graph

The derivative of f at x is the function derived from . . .

Give the formal definition of the **derivative.**

The derivative $f'(x)$ is a formula for . . .

Example 2: Find the derivative of $f(x) = 9 - 2x^2$.

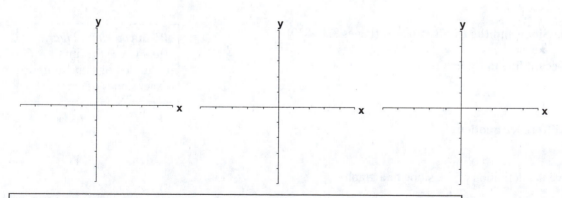

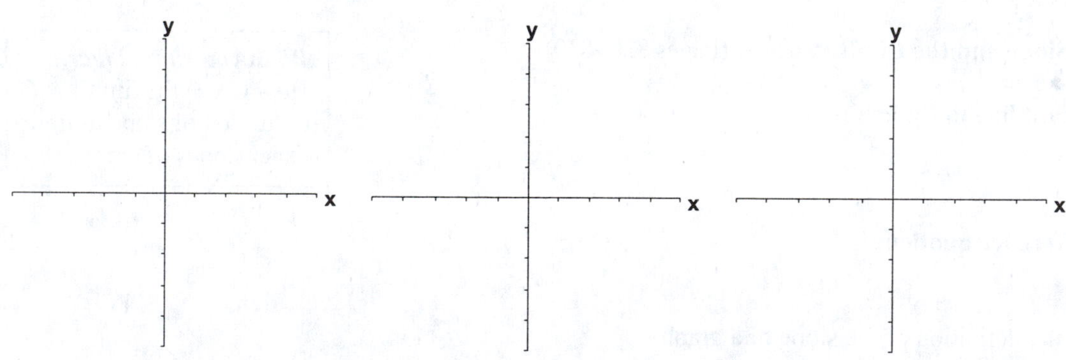

Homework Assignment

Page(s)

Exercises

Course Number

Instructor

Date

Section 12.4 Limits at Infinity and Limits of Sequences

Objective: In this lesson you learned how to evaluate limits at infinity and find limits of sequences.

What you should learn
How to evaluate limits of functions at infinity

I. Limits at Infinity and Horizontal Asymptotes
(Pages 835–838)

Define **limits at infinity.**

To help evaluate limits at infinity, you can use the following:

If r is a positive real number, then $\lim_{x\to\infty} \frac{1}{x^r} =$ ________ .

If x^r is defined when $x < 0$, then $\lim_{x\to-\infty} \frac{1}{x^r} =$ ________ .

Example 1: Find the limit: $\lim_{x\to\infty} \frac{1+5x-3x^3}{x^3}$

If $f(x)$ is a rational function and the limit of f is taken as x approaches ∞ or $-\infty$,

- When the degree of the numerator is less than the degree of the denominator, the limit is ________ .
- When the degrees of the numerator and the denominator are equal, the limit is ________ .
- When the degree of the numerator is greater than the degree of the denominator, the limit ________ .

Larson/Hostetler/Edwards *Precalculus Functions and Graphs: A Graphing Approach, Third Edition*
Larson/Hostetler/Edwards *Precalculus with Limits: A Graphing Approach, Third Edition*
Student Success Organizer

II. Limits of Sequences (Pages 839–840)

What you should learn
How to find limits of sequences

For a sequence whose nth term is a_n, as n increases without bound, if the terms of the sequence get closer and closer to a particular value L, then the sequence is said to ______________ to L. Otherwise, the sequence is said to ______________.

Give the definition of the limit of a sequence.

Example 2: Find the limit of the sequence $a_n = \dfrac{(n-3)(4n-1)}{4-3n-n^2}$.

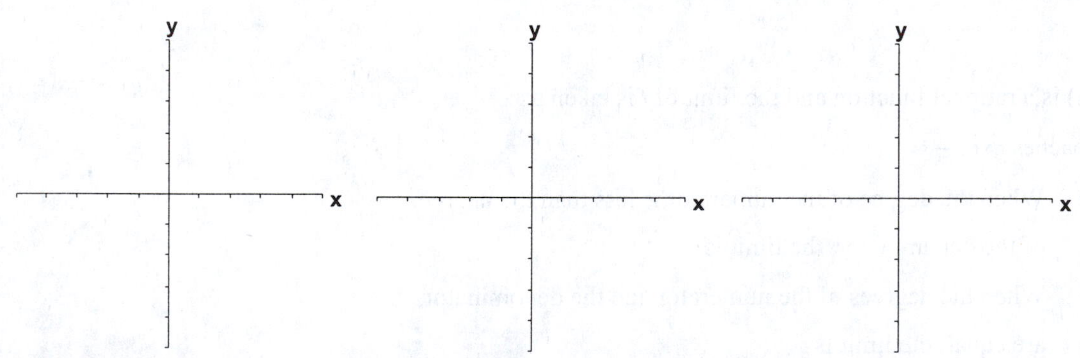

Homework Assignment

Page(s)

Exercises

Course Number

Instructor

Date

Section 12.5 The Area Problem

Objective: In this lesson you learned how to find limits of summations and use them to find areas of regions bounded by graphs of functions.

I. Limits of Summations (Pages 843–845)

What you should learn
How to find limits of summations

The following summation formulas and properties are used to evaluate finite and infinite summations.

1. $\sum_{i=1}^{n} c =$

2. $\sum_{i=1}^{n} i =$

3. $\sum_{i=1}^{n} i^2 =$

4. $\sum_{i=1}^{n} i^3 =$

5. $\sum_{i=1}^{n} (a_i \pm b_i) = \sum_{i=1}^{n} a_i \pm \sum_{i=1}^{n} b_i$

6. $\sum_{i=1}^{n} ka_i = k\sum_{i=1}^{n} a_i$

To find the limit of a summation, . . .

Example 1: Find the limit of $S(n)$ as $n \to \infty$.

$$S(n) = \sum_{i=1}^{n} \frac{i-5}{n^3}$$

II. The Area Problem (Pages 846–848)

What you should learn
How to use rectangles to approximate areas of plane regions and to use limits of summations to find areas of plane regions

Describe the area problem.

The exact area of a plane region R is given by . . .

Let f be continuous and nonnegative on the interval $[a, b]$. The area A of the region bounded by the graph of f, the x-axis, and the vertical lines $x = a$ and $x = b$ is

Example 1: Find the area of the region bounded by the graph of $f(x) = (x-4)^2 + 5$ and the x-axis between $x = 3$ and $x = 6$.

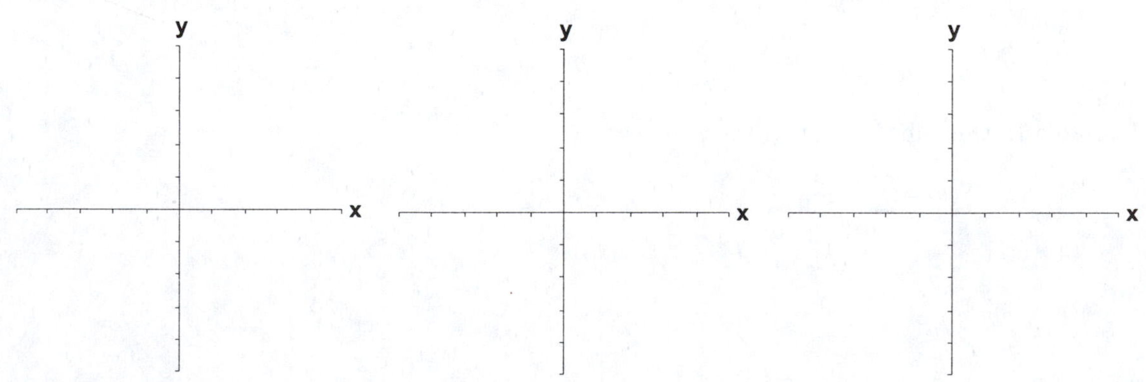

Homework Assignment

Page(s)

Exercises